Pharmaceutical Discovery, Development and Manufacturing Forum

Held at the 2024 AIChE Annual Meeting

San Diego, California, USA
27-31 October 2024

ISBN: 979-8-3313-1662-4

Pharmaceutical Discovery, Development and Manufacturing Forum

Held at the 2024 AIChE Annual Meeting

San Diego, California, USA
27-31 October 2024

AIChE
The Global Home of Chemical Engineers

Printed from e-media with permission by:

Curran Associates, Inc.
57 Morehouse Lane
Red Hook, NY 12571

Some format issues inherent in the e-media version may also appear in this print version.

Printed with permission by Curran Associates, Inc. (2025)

AIChE
120 Wall Street, FL 23
New York, NY 10005-4020

Phone: (800) 242-4363
Fax: (203) 775-5177

www.aiche.org

Additional copies of this publication are available from:

Curran Associates, Inc.
57 Morehouse Lane
Red Hook, NY 12571 USA
Phone: 845-758-0400
Fax: 845-758-2633
Email: curran@proceedings.com
Web: www.proceedings.com

TABLE OF CONTENTS

DIVISION PLENARY: PHARMACEUTICAL DISCOVERY DEVELOPMENT AND MANUFACTURING (INVITED TALKS)

CONTINUOUS PROCESSING IN DRUG PRODUCT

ENABLING TECHNOLOGIES RELEVANT TO DRUG SUBSTANCE

ADVANCES IN CONTROL STRATEGY FOR CONTINUOUS PROCESSES AND UTILIZATION OF PATS

ADVANCEMENTS IN DRUG PRODUCT PARTICLE ENGINEERING AND MATERIAL SCIENCE

PREDICTIVE SCALE-UP/SCALE-DOWN FOR PHARMACEUTICALS

PROCESS MODELS FOR DRUG SUBSTANCE, DRUG PRODUCT, AND BIOPHARMACEUTICALS PART 1

ADVANCES IN DRUG DISCOVERY PROCESSES (INCLUDING HTE): ADVANCED TECHNOLOGY APPROACHES TO MAXIMIZE PUBLIC HEALTH IMPACTS

ADVANCES IN NEW MODALITIES: NEW MODALITIES AND TECHNOLOGIES

PHARMA 4.0 (ADVANCED CONTROLS, PROCESS AUTOMATION, DATA ANALYTICS, ETC.)

POSTER SESSION: PHARMACEUTICAL DISCOVERY, DEVELOPMENT, AND MANUFACTURING

INVITED TALKS: SUSTAINABILITY IN THE PHARMACEUTICAL INDUSTRY

ADVANCES IN DRUG DISCOVERY PROCESSES (INCLUDING HTE): PROTEIN ENGINEERING APPROACHES WITH TARGET THERAPEUTIC APPLICATIONS

ADVANCES IN NEW MODALITIES: SEPARATIONS AND FORMULATION

ENABLING TECHNOLOGIES RELEVANT TO DRUG PRODUCT

PREDICTIVE SCALE-UP/SCALE-DOWN FOR BIOPHARMACEUTICALS

ADVANCES IN CONTROL STRATEGY: APPLICATIONS TO BIOPHARMACEUTICALS APPLICATIONS

ADVANCES IN CONTROL STRATEGY TOWARDS DRUG SUBSTANCE AND DRUG PRODUCT UNIT OPERATIONS

ADVANCES IN CONTROL STRATEGY USING MODELING TOOLS AND APPROACHES

ADVANCEMENTS IN DRUG SUBSTANCE PROCESSING AND MATERIAL SCIENCE

CONTINUOUS PROCESSING IN DRUG SUBSTANCE

CONTINUOUS PROCESS: NEXT GENERATION ADVANCEMENTS

INVITED TALKS: IN HONOR OF RAMKI'S 85TH BIRTHDAY TO COMMEMORATE HIS CONTRIBUTIONS TO BIOPHARMACEUTICAL INDUSTRY

PROCESS MODELS FOR DRUG SUBSTANCE, DRUG PRODUCT, AND BIOPHARMACEUTICALS PART 2

DATA-DRIVEN APPROACHES, AND ML/AI FOR PHARMACEUTICAL APPLICATIONS

COMPUTATIONAL SOLID STATE PHARMACEUTICS

INVITED TALKS: IN HONOR OF PROF. ROBERT PRUD'HOMME

490c Combining Mechanistic Models and Data to Aid Pharmaceutical Process Development

Kimberley B. McAuley

Kim McAuley is Professor of Chemical Engineering at Queen's University with a PhD from McMaster University. Her research focuses on the development and application of first-principles models of chemical processes. Kim McAuley develops practical statistical methods for combining models with data, leading to reliable models for process scale-up, optimization, and control. Her techniques are widely used by academic and industrial researchers. She serves on the Board of Directors for *International Society for Chemical Reaction Engineering* and International Advisory Boards for three journals. She received the *DG Fisher Award* from the Systems & Control Division of the Canadian Society for Chemical Engineering, a *Queen Elizabeth II Diamond Jubilee Medal* from the Government of Canada, a *MITACS Mentorship Award of Excellence*, and a Queen's University's *Award for Excellence in Graduate Supervision.* She is a Fellow of the Canadian Academy of Engineering and a Fellow of the Chemical Institute of Canada.

490a Developing a Novel Platform for Synthetic Peptide Manufacture Using Continuous Synthesis Methodologies

Kevin Seibert

Kevin is a technical VP/Senior Fellow with Lilly and holds a PhD in Chemical Engineering. He has over 25 years' experience working in small and medium molecule syntheses for both Merck and Eli Lilly. Kevin has been responsible for the process development and optimization of many early as well as late phase products and led the technology transfer, validation, and launch of several marketed compounds. Kevin has also led initiatives for developing enhanced filings in support of the ICH QbD and continuous processing initiatives. Most recently, Kevin led the late phase development team responsible for the commercialization of Lilly's Mounjaro. Kevin has worked closely with expert teams within Lilly to implement a strategy for synthesizing peptides using a standardized manufacturing platform and applying the strategy to Lilly's growing peptide portfolio.

Currently Kevin is a Fellow with the American Institute of Chemical Engineers, and Chair of the PD2M forum within AIChE, driving the application of chemical engineering principles to discovery, development and manufacturing within the pharmaceutical industry.

490b Molecular Engineering for Pharmaceutical Process Development

Claire S. Adjiman

Claire Adjiman is Professor of Chemical Engineering at Imperial College London. She holds an MEng from Imperial College and a PhD from Princeton University, both in Chemical Engineering. Her research is focused on multiscale process and molecular/materials design, including the development of design methods, property prediction techniques and optimisation algorithms. She works extensively with industry, especially the pharmaceuticals, agrochemicals and energy sectors and has licensed thermodynamic modelling software.

She is a Fellow of the Royal Academy of Engineering (2015), an International Honorary Member of the American Academy of Arts and Sciences (2022) and of the US National Academy of Engineering (2023). She is also a Fellow of the Institution of Chemical Engineers and the Royal Society of Chemistry. She has received awards that include a RAEng-ICI Fellowship (1998-2003), the Philip Leverhulme Prize for Engineering (2009), the SCI Armstrong Lecture (2011), an EPSRC Leadership Fellowship (2012-2017), and the RSC Elizabeth Colbourn Memorial Lecture (2020), the American Institute of Chemical Engineers' Computing in Chemical Engineering Award (2021) and the George Stephanopoulos Award for Process Systems Engineering (2024). She is Editor-in-Chief of *Molecular Systems Design and Engineering,* and she is a member of the editorial boards of *Computers and Chemical Engineering* and *Fluid Phase Equilibria*. At Imperial, she was a Founding Co-Director of the Institute for Molecular Science and Engineering (2015-2020) and she is Director of the Sargent Centre for Process Systems Engineering. She is a Trustee of Future Innovation in Process Systems Engineering (FIPSE). She is/has been a member of several advisory bodies, in the UK (EPSRC Strategic Advisory Network) and Singapore (CARES-C4T), and chairs the Scientific Advisory Board of the Max Planck Institute for Dynamics of Complex Technical Systems Magdeburg (Germany).

550a A Modular Continuous Pharmaceutical Injectable Manufacturing Pilot-Plant and a Novel Digital Twin

Ravendra Singh

Advanced continuous manufacturing (CM) is emerging as a preferred platform to produce active pharmaceutical ingredients (APIs) as well as full finished injectable liquid dosage forms [1]. The future continuous pharmaceutical manufacturing is desired to be based on modular plug and play concept. In this approach, mobile modules consisting of a group of unit operations have been developed that can be easily reconfigured to assemble a scalable pharmaceutical manufacturing process for different products. It is also suitable for on-demand manufacturing such as in the case of pandemic. However, because of different levels of complexity, the development and adaptation of flexible and modular CM plant for injectable manufacturing is still an open area of research. Therefore, a systematic framework of methods and tools are needed to develop, adapt, and evaluate the flexible modular CM plant for pharmaceutical manufacturing of liquid dosage forms with reduced time and resources. In particular, a digital twin is critical for quick design, adaptation, optimization, and control of manufacturing process, allowing for virtual modeling of operations without extensive process development [1-2].

In this work, a modular pilot-plant and digital twin model of continuous manufacturing of full finish final liquid dosage forms has been developed. The developed continuous injectable manufacturing process takes the API synthesized and purified via a continuous API manufacturing process, either in powder form or as a solution in the final liquid ingredient of the formulation and turns it into a finished liquid product. This process is composed of three modules. In module 1, the API is conditioned in a preconditioning tank. This involves either dissolving powder API, or diluting an API liquid feedstock, to generate a solution with the desired concentration. Solutions of all remaining formulation ingredients are dispensed from additional refillable tanks. All ingredients are pumped at controlled mass flow rates to module 2. The module 2 consist of a static mixer, a homogenizer, and an ultrafiltration unit operation. The ultrafiltration membrane is used for purification and sterilization. In module 3, the outlet stream from ultrafiltration unit is pumped into a reservoir tank (surge capacity), from where it is metered-fed into vial filling machine. The filled vials are then transferred to a capping station.

The developed digital twin model library consists of the mathematical model of unit operations involved in modules 1-3. In module 1, the continuous feeding system model including feed tanks model, refill system model, and pump model have been developed. In module 2, the mathematical model of static mixer to predict the mixing of different ingredients, a homogenizer model to predict the agglomerate size reduction, and an ultrafiltration model to predict the different critical process parameters (CPPs) including retentate and permeate fluxes have been developed. In module 3, the surge tank model has been developed to make the balance between modules 2 and 3. The vial filling model has been developed to predict the final critical quality attributes (CQAs) of the injectable drug product. The model is predictive in nature and has been verified using the experimental data generated from our injectable CM pilot-plant.

The objective of this presentation is two-fold; first to highlight the developed unit operation model library for the continuous pharmaceutical manufacturing of liquid dosage forms and then demonstrate its application for integrated flowsheet model generation, dynamic optimization, and design of control system.

References

1. O'Connor, T. (2020). FDA Grand Rounds: Modernization of Pharmaceutical Manufacturing through the Adoption of Advanced Technology. https://www.fda.gov/science-research/fda-grand-rounds/fda-grand-rounds-modernization-pharmaceutical-manufacturing-through-adoption-advanced-technology.
2. Singh, R. (2023). Advanced pharmaceutical continuous manufacturing of liquid dosage forms. Pharma Focus Europe. Issue 3, 55-59. https://www.pharmafocuseurope.com/manufacturing/advanced-pharmaceutical-continuous-manufacturing
3. Singh, R., Sahay, A., Fernando Muzzio, Ierapetritou, M., Ramachandran, R. (2014). A systematic framework for onsite design and implementation of the control system in continuous tablet manufacturing process. Computers & Chemical Engineering Journal, 66, 186-200.

Acknowledgements

This work is supported by the US Food and Drug Administration (FDA) under contract number 75F40122C00122.

550b Model-Based Design and Control of a Platform for Continuous Production of Recombinant Adeno-Associated Virus

Francesco Destro, John Joseph, Caleb Neufeld, Jacqueline M. Wolfrum, Paul W. Barone, Stacy Springs, Anthony J. Sinskey, Richard D. Braatz

The demand for recombinant adeno-associated virus (rAAV) is poised to exceed current production capacities, driven by the increasing number of approved and advanced-phase rAAV-based gene therapies [1]. Current rAAV manufacturing processes yield less-than-optimal titers, consisting of a significant proportion of non-functional empty capsids [2]. FDA recently approved the first rAAV-based gene therapies manufactured in the Sf9 insect cell line with the baculovirus expression vector system (BEVS). Within the Sf9/BEVS platform, Sf9 cells produce rAAV as a result of infection with recombinant baculoviruses that carry the genetic blueprint for rAAV vector production. Transitioning from the traditional batch manufacturing to continuous processing would allow to significantly increase the BEVS productivity. However, the formation of baculovirus defective interfering particles severely compromises the process efficiency and leads to shutdown of continuous operation in the BEVS within, at best, 5–7 days from viral inoculation [3]. In this work, a mathematical model is developed and used for in silico design of a process for continuous rAAV manufacturing in the BEVS [4,5]. The mathematical model is used to assess the optimal operating conditions of cell growth and production bioreactors that can counteract the detrimental effect of baculovirus defective interfering particles. The process designed through the mathematical model is implemented and validated on laboratory-scale bioreactors, achieving for the first time continuous production of high titers of rAAV for over 20 days. This study paves the way for efficient and scalable continuous rAAV manufacturing, positioning the industry to meet the rising demand for gene therapies.

References

[1] Au, H.K.E., Isalan, M. and Mielcarek, M., 2022. Gene therapy advances: a meta-analysis of AAV usage in clinical settings. *Front. Med.* **8**, 809118.

[2] Su, W., Patrício, M.I., Duffy, M.R., Krakowiak, J.M., Seymour, L.W. and Cawood, R. (2022). Self-attenuating adenovirus enables production of recombinant adeno-associated virus for high manufacturing yield without contamination. *Nat. Commun.* **13**(1), 1182.

[3] Vignuzzi M. and López, C. B (2019). Defective viral genomes are key drivers of the virus-host interaction. *Nat. Microbiol.* **4**, 1075–1087.

[4] Destro, F. and Braatz, R.D. (2024). Efficient simulation of viral transduction and propagation for biomanufacturing. bioRxiv, DOI: 10.1101/2024.03.30.587435.

[5] Destro, F., Joseph, J., Srinivasan, P., Kanter, J.M., Neufeld, C., Wolfrum, J.M., Barone, P.W., Springs, S.L., Sinskey, A.J., Cecchini, S., Kotin, R.M., and Braatz, R.D. (2023). Mechanistic modeling explains the production dynamics of recombinant adeno-associated virus with the baculovirus expression vector system. *Mol. Ther. Methods Clin. Dev.* **30**, 122-146.

550c Path to Commercial Reality of Continuous Manufacturing - a Cdmo Perspective

Jose Luis Santos

Continuous tableting (CT, defined as continuous manufacturing of oral dosage tablets) represents a paradigm shift in pharmaceutical manufacturing, promising increased efficiency, reduced costs, and accelerated pathways for drug approval. The end users of CT, and especially the CDMOs, play an important role in the adoption of this technology. However, there remain both tangible and perceived obstacles that must be addressed to facilitate widespread adoption.

This session will cover progress being made across multiple fronts and the roadmap towards commercial adoption of the technology.

Details will be provided on engaging with a complex operation involving a large team, challenging sampling procedures, and complex logistics in the process of launching continuous tableting GMP operations. Each step, from running the rig to changeover, presented unique and complex challenges necessitating innovative solutions. The transition from a team of experts to a proficient commercial operation run by operators will be highlighted. Additionally, the journey to a proficient commercial operation, with improved yields and reduced deviations, will be reviewed. The specific challenge of changeover will be discussed, with examples on how it was addressed through simulation modeling and shop-floor operational excellence approaches.

The economics of CT technology will be explored, with insights into its cost competitiveness with batch mode. Based on cost analysis, areas requiring attention will be discussed, such as team size, equipment costs, and maintenance.

The discussion will also highlight recent regulatory guidance and publications, exploring the trend towards accelerated pathways for drug approval, with continuous tableting emerging as a facilitator of acceleration. Realizing the benefits of CT in accelerating drug approval processes requires simplification and flexibility, aligning with regulatory expectations and industry standards.

Attention will be given to the development of simple and flexible CT standards, enabling new modes of operation and ultimately simplifying CT technology through the introduction of flexible rigs. The discussion will include details of the CDC Flex concept, which covers a range of line rates from low to high and encompassing both batch and continuous modes of operation.

Lastly, the session will include a discussion on the journey of CT from complexity to commercial reality, and the diverse opportunities for further simplification and advancement of the technology that result from Hovione's experience as early adopters in this field.

550d Methodology for Rapid Development of a Continuous Loss-in-Weight Feeding Process

Ana Martins, Claudia Moura, João Lopes

The pharmaceutical industry has been increasingly shifting towards continuous manufacturing, specifically for the production of tablets. When compared to batch processing, continuous tableting exhibits higher efficiency, flexibility and product quality. The first unit operation in a continuous tableting process is the feeding step, where each raw material is fed through a loss in weight (LIW) feeder at a predefined rate. In order to ensure the quality of the final product, it is crucial to maintain an accurate and stable mass flow, making it essential to understand the behavior of the individual raw materials during the feeding process [1, 2]. Despite the available research correlating material properties with feeding performance, the implementation of this knowledge is not straightforward due to the diversity and inherent variability of raw materials, equipment, external factors, and modeling strategies [3]. Additionally, there is still a lack of understanding of how different setup options and process parameters affect the feeding performance, thus making it challenging to use the models already available.

The present work focuses on creating a methodology for rapid development of a continuous feeding process by creating predictive models that correlate raw material properties with their feeding performance. Additionally, the impact of various setup options and process parameters was assessed, enabling the creation of a workflow to predict the operating space and ideal setup for a new material based solely on material properties.

To obtain a broad range of material properties, 18 raw materials were selected: nine commercially available excipients, two model APIs, and seven processed excipients. The materials were then submitted to various characterization techniques, such as bulk and tapped density, true density, shear cell (yield locus and wall yield locus test), FT4 (stability and variable flow rate, permeability, aeration, and compressibility tests), angle of repose, particle size distribution, dynamic vapor sorption, and Faraday Cup charge measurement. In total, 63 material property variables were obtained, describing density, flowability, cohesion, particle size, hygroscopicity, and electrostatic charge. Each material was tested in a GEA compact feeder using a predefined setup and feeding parameters, to obtain the feeding performance responses (i.e., mass flow RSD, maximum feeding capacity, feed factor decay constant, and dead mass). Multivariate data analysis (MVDA) was used to create predictive models correlating raw materials properties with feeding performance parameters. Due to the considerable number of characterization variables, careful attention was taken during the analysis to avoid model overfitting. To evaluate the impact of altering the setup and feeding parameters on the feeding responses, a design of experiments (DoE) was performed by changing the screw pitch, mesh, agitator, baffle and screw speed.

Results from the MVDA show that feeding performance is mainly affected by variables related to density, cohesion, and flowability. For instance, materials with higher density and lower cohesion have lower mass flow RSD, as well as a higher maximum feeding capacity (FFmax), due to a more uniform screw filling. On the contrary, more cohesive materials led to an earlier decay of the feed factor (FFdecay constant) since they will gradually stick to the screws, reducing the free screw volume over time. Additionally, these materials also tend to stick to the hopper walls, leading to a higher dead mass.

Results from the DoE within the tested ranges show that the mass flow RSD is higher when running at higher screw speeds, as well as when using a mesh or the asymmetric agitator. For more cohesive materials, the asymmetric agitator also leads to an increase in dead mass due to the accumulation of the material on the hopper walls caused by the two vertical shafts. This accumulation can lead to powder falling from the walls over time, causing spikes in the feeder weighing scale and thus affecting the mass flow variability. The FFmax is only affected by the screw pitch, and no impact is observed when using the vertical baffle.

The predictive models created and the conclusions from the DoE were incorporated into a workflow to aid in process development. Since the models have an associated error, this workflow was validated with five external materials by predicting their feeding performance and ideal setup (including refill strategy) using only raw material characterization data. Despite these errors, the selected setup was the same for the experimental and predicted feeding performance. Overall, in order to predict the feeding apparatus and performance, only four characterization techniques are deemed relevant, reducing the analysis time from two days to one hour. As a result, the quantity of material needed to make predictions was also significantly reduced.

Although the implementation of this workflow does not eliminate the need to perform laboratory feeding trials before transitioning to a continuous tableting line, it serves as a fit-for-purpose tool that enables the reduction of raw material and time spent in development. Additionally, it provides deeper process knowledge, enabling the transition from a trial-and-error approach to a data-driven approach.

References:

[1] Wang, Y., Li, T., Muzzio, F. J., & Glasser, B. J. (2017). Predicting feeder performance based on material flow properties. *Powder Technology*, 308, 135–148. https://doi.org/10.1016/j.powtec.2016.12.010

[2] Van Snick, B., Holman, J., Cunningham, C., Kumar, A., Vercruysse, J., De Beer, T., Remon, J. P., & Vervaet, C. (2017). Continuous direct compression as manufacturing platform for sustained release tablets. *International Journal of Pharmaceutics*, 519(1–2), 390–407. https://doi.org/10.1016/j.ijpharm.2017.01.010

[3] Hörmann-Kincses, T., Beretta, M., Kruisz, J., Stauffer, F. J., Birk, G., Piccione, P. M., Holman, J., & Khinast, J. G. (2022). Predicting powder feedability: A workflow for assessing the risk of flow stagnation and defining the operating space for different powder-feeder combinations. *International Journal of Pharmaceutics*, 629, 122364. https://doi.org/10.1016/j.ijpharm.2022.122364

Abstract Title: Model-Based (Digital Twins) Development for Continuous Manufacturing of mRNA (messenger Ribonucleic Acid) Vaccines and RNA-Based Therapeutics

The global urgency prompted by the SARS-CoV-2 (COVID-19) pandemic accelerated the development of messenger ribonucleic acid (mRNA) vaccines, which have proven to be a transformative technology in the fight against infectious diseases. Beyond COVID-19, mRNA vaccines have ignited renewed interest in RNA-based therapeutics, with applications now being actively explored in diverse fields, including oncology, metabolic disorders, neurological diseases, and genetic disorders. Researchers and biopharmaceutical companies are leveraging the potential of RNA technologies, expanding the horizon for future RNA-based treatments. Additionally, the pandemic has underscored the importance of enhancing society's preparedness for future public health crises, leading to a strategic focus on optimizing the development and production pipelines for RNA technologies to ensure swift and effective responses.

Despite these advancements, RNA manufacturing presents several key challenges, such as issues related to production scalability, maintaining high-quality output, ensuring the efficacy and safety of therapeutics, and addressing the high costs largely driven by RNA's inherent instability. These challenges are compounded by inefficiencies in traditional manufacturing approaches, which hinder the rapid production of RNA-based therapeutics at scale.

To address these challenges, the biopharmaceutical industry is recognizing the need to revolutionize traditional RNA manufacturing processes. Central to this transformation is the integration of upstream and downstream manufacturing operations into a continuous, seamless production system. Continuous manufacturing offers the potential to not only meet the growing demand for RNA-based therapeutics but also to enhance the resilience of the drug supply chain by reducing contamination risks and improving production efficiency. By significantly reducing time to market, continuous manufacturing could play a pivotal role in ensuring rapid and scalable responses to both current and future healthcare needs.

A key enabler of this shift towards continuous manufacturing is the use of advanced modeling and simulation techniques. These techniques are now being endorsed by industry leaders, policymakers, and regulatory agencies such as the FDA and EMA, who recognize the need for precise and robust process control in RNA manufacturing. Digital twin technology—real-time, virtual representations of physical processes—offers a powerful tool for optimizing every stage of RNA vaccine and therapeutic production. Digital twins allow for real-time monitoring and control of unit operations, ensuring that production processes are continuously optimized for efficiency, quality, and safety.

This work focuses on two critical dimensions of continuous RNA manufacturing:

1. **Application of Modeling and Simulation Techniques**: The first dimension involves the application of modeling and simulation to dissect and understand the intricacies of unit operations within a continuous RNA manufacturing framework. The use of digital twins enables real-time data integration, allowing for dynamic adjustments to production parameters to maintain optimal conditions for RNA stability, yield, and purity. This approach facilitates a deeper understanding of process interdependencies, ensuring that each unit

operation—from mRNA synthesis to downstream purification—is fine-tuned for maximum efficiency.

2. **Validation of Models with Experimental Data**: The second dimension of our research is the rigorous validation of these models through extensive comparison with experimental data. This step is crucial for ensuring that the models accurately reflect real-world manufacturing conditions. Validation not only confirms the robustness of the models but also enhances confidence in their ability to predict performance under varying conditions. By achieving high predictive accuracy, digital twins can reduce the need for costly and time-consuming physical trials, accelerating the development and approval timelines for new RNA-based therapies.

Our research aims to bridge the gap between theoretical model development and practical application in RNA manufacturing. By integrating upstream and downstream operations and validating these processes with real-world data, we are paving the way for a new era of continuous RNA production. This approach holds the promise of streamlined manufacturing processes, reduced costs, enhanced product quality, and faster time-to-market for RNA-based therapeutics. Moreover, it positions the biopharmaceutical industry to respond more effectively to global healthcare challenges, ensuring that innovative therapies can be produced quickly and at scale when needed most.

In conclusion, the adoption of model-based development and digital twin technology represents a significant step forward in the manufacturing of RNA vaccines and therapeutics. By embracing continuous manufacturing and leveraging the power of simulation and real-time data integration, the industry can overcome the many challenges that have traditionally hindered RNA production. This research is poised to play a critical role in shaping the future of RNA-based therapeutics, bringing about a new level of efficiency, safety, and scalability in drug manufacturing.

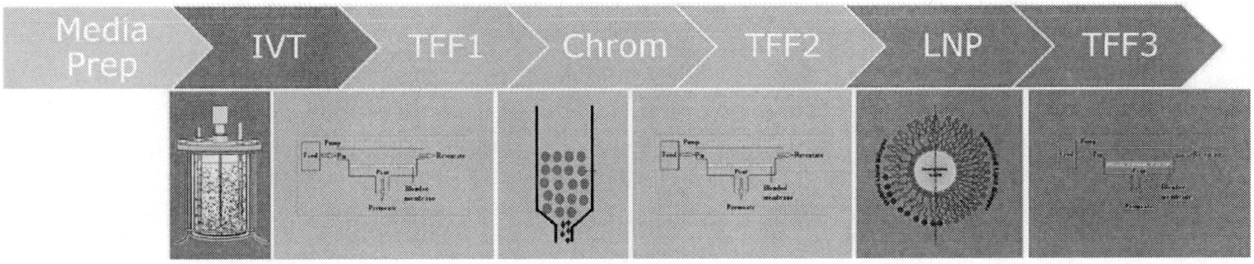

550f Development of a Continuous Rotary Filtration System in Advancing the Continuous Manufacturing of Pharmaceuticals

Wei Wu, Shalabh Yeole, Chuntian Hu

As pharmaceutical manufacturers are looking for ways to shift toward continuous manufacturing, one of the biggest challenges has been the continuous separation and purification of active pharmaceutical ingredients (APIs) or intermediates after crystallization or reactive crystallization. A significant gap exists between continuous crystallization and the availability of corresponding filtration equipment[1]. Keeping this obstacle in mind, a continuous rotary filter was developed, and its design space was investigated using a commercially available API[2]. The filtration unit consists of a rotatable plate that receives a continuous flow of feed material. The feed was then evenly distributed along the radius of the plate, forming a thin cake. The cake is continuously filtered, washed, and removed from the plate. An additional plate wash continually cleans the filter medium, enabling long-term operations. On the stand-alone unit, a slurry containing the selected API and two main impurities was processed. Process parameters including cake residence times and cake wash rates were investigated. By operating the filter continuously, the process can be adjusted to minimize the cake resistance, therefore facilitating efficient filtration performance. Process parameters for traditional filtration study such as cake resistance and filtration time became constant throughout the continuous operation. The continuous filter was then incorporated into an integrated continuous manufacturing (ICM) line at pilot scale, where it processed the slurry from a crystallizer and then transferred the wet cake into the downstream process[3]. In both settings, the rotary filter isolated and purified the cake to the required quality specifications, proving its effectiveness and robustness during continuous operations. The process design has been further improved to allow technology transfer for GMP manufacturing. An advanced filter unit operation is designed with GMP compliant approaches, and the cleaning-in-place (CIP) capacity is incorporated. We report the commissioning of a pilot-scale filter with high containment capabilities, which can be used in laboratory or piloting environments to develop manufacturing technologies prior to commercial implementation. This advancement expands R&D activities for continuous manufacturing to include potent compounds isolations and purification, enabling a broader scope for pharmaceutical development.

References

(1) Cote, A.; Erdemir, D.; Girard, K. P.; Green, D. A.; Lovette, M. A.; Sirota, E.; Nere, N. K. Perspectives on the Current State, Challenges, and Opportunities in Pharmaceutical Crystallization Process Development. *Cryst Growth Des* **2020**, *20* (12), 7568–7581. https://doi.org/10.1021/acs.cgd.0c00847.

(2) Wu, W.; Sayin, R.; Shvedova, K.; Born, S. C.; Testa, C. J.; Yeole, S. S.; Censullo, A. S.; Srivastava, A. K.; Ramnath, A.; Hu, C.; Takizawa, B.; O'Connor, T. F.; Yang, X. B.; Ramanujam, S.; Mascia, S. A Continuous Rotary Filtration for the Separation and Purification of an Active Pharmaceutical Ingredient. *Org Process Res Dev* **2023**. https://doi.org/10.1021/acs.oprd.3c00263.

(3) Testa, C. J.; Hu, C.; Shvedova, K.; Wu, W.; Sayin, R.; Casati, F.; Halkude, B. S.; Hermant, P.; Shen, D. E.; Ramnath, A.; Su, Q.; Born, S. C.; Takizawa, B.; Chattopadhyay, S.; O'Connor, T. F.; Yang, X.; Ramanujam, S.; Mascia, S. Design and Commercialization of an End-to-End Continuous Pharmaceutical Production Process: A Pilot Plant Case Study. *Org Process Res Dev* **2020**, *24* (12), 2874–2889. https://doi.org/10.1021/acs.oprd.0c00383.

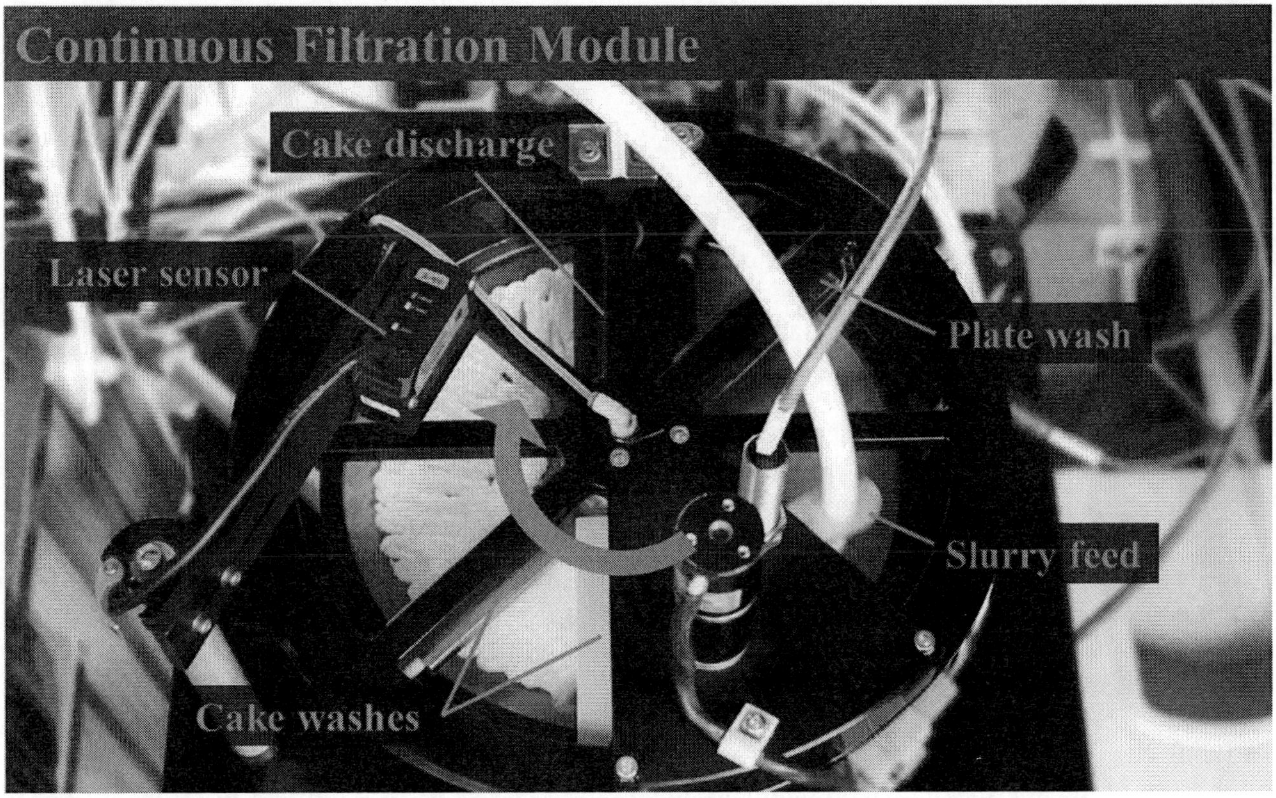
Continuous Filtration Module
Cake discharge
Laser sensor
Plate wash
Slurry feed
Cake washes

550g Navigating API Particle Morphology Challenges to Streamline Continuous Drug Product Manufacturing

Eftychios (Efty) Hadjittofis, Nathan de Bruyn, Adam Hill, Oleksii Shemchuk, Jerome Mantanus

In the pioneering work "On the Equilibrium of Heterogeneous Substances", Prof. J.W. Gibbs, provided, for the first time, a description of the general rules determining the morphology of a growing crystal stating that: "The shape of a crystal or the form of a crystal is determined by the mode of growth", while he, already notes that " In many cases, the shape of a particle is intimately connected with its function or utility in a particular application or process.'

Nowadays, 150 years after the publication of Gibbs' seminal work, the reconciliation of crystallisation, a process with dual purification and particle engineering function, with downstream, drug product manufacturing processes, becomes increasingly important, in the biopharmaceutical industry. The need for continuous and dry drug product (DP) processing gets momentum, thanks to business and scientific considerations. The hitherto published literature, along with the accumulated experience within the community, highlights that having Active Pharmaceutical Ingredient (API) particles with the right particle characteristics is paramount for the development of any continuous dry DP process operation.

Targeting the desired particle morphology, with an industrially relevant way, stands as out as a conundrum in the pharmaceutical crystallisation community. A growing number of APIs appears to exhibit persistent needle behaviour. Identifying early on, during the development of an API, the potential morphologies, the API can assume, and the available particle morphology levers emerges as a critical capability for API isolation process development and de-risking.

UCB's crystallisation development workflow addresses this challenge, through the CrystalGrower platform. The deployment of this in-silico tool allowed us to screen the morphologies of our APIs at different crystal growth conditions (solvents, temperatures, supersaturation profiles). As the main input is only the crystal structure of the API, this tool can be deployed early in the development workflow. This is quite groundbreaking, it allows the project team to decide, from the Ph I of the clinical development, the most suitable particle engineering strategy, identify crystallisation solvents of particular interest for particle morphology control and even to predict whether secondary particle modifications are necessary to tailor the particle properties. In this framework, the particle morphology toolbox synergises with other tools from the crystallisation development workflow, such as the solubility prediction app and miniaturised screening platforms, such as the Crystalline. The results to be presented come from two on-going therapeutic projects demonstrating the flexibility gained by the introduction of a robust particle morphology prediction framework.

344a Sequential Flash Nanoprecipitation for the Scalable Formulation of Stable Core-Shell Nanoparticles with Core Loadings up to 90%

Nicholas Caggiano, Satya Nayagam, Leon Z. Wang, Brian K. Wilson, Parker Lewis, Shadman-As-Sami Jahangir, Rodney Priestley, Robert K. Prud'homme, Kurt Ristroph

Flash NanoPrecipitation (FNP) is a scalable, single-step process that uses rapid mixing to formulate nanoparticles with a hydrophobic core and amphiphilic stabilizing shell. Because the two steps of particle self-assembly – (1) core nucleation and growth and (2) adsorption of a stabilizing polymer onto the surface of the growing core – occur simultaneously during FNP, nanoparticles formulated with a weight fraction of core material above approximately 70% typically exhibit poor stability or do not form at all. In addition, a fundamental limit on the concentration of total solids that can be introduced into the FNP process has been reported previously. Both of these limits (on core loading and total input solids concentration) are believed to share a common mechanism: the entrainment of the stabilizing polymer into the growing particle core, which leads to destabilization and aggregation.

We here demonstrate a variation of the FNP technique that separates the nucleation and stabilization steps of particle formation into separate sequential mixers. This scheme allows the hydrophobic active to nucleate and grow in the first mixing chamber unimpeded by adsorption of the stabilizing polymer, which is not introduced until the growing nuclei enter the second mixer immediately downstream of the first. Using this Sequential Flash NanoPrecipitation (SNaP) technique, we formulate stable nanoparticles with up to 90% core loading by mass. We also form particles at 6-fold higher total input solids concentrations than typically reported. This approach of decoupling the nucleation and stabilization of particles produced by FNP broadens the range of loadings and solids concentrations that can be achieved by the platform without sacrificing scalability.

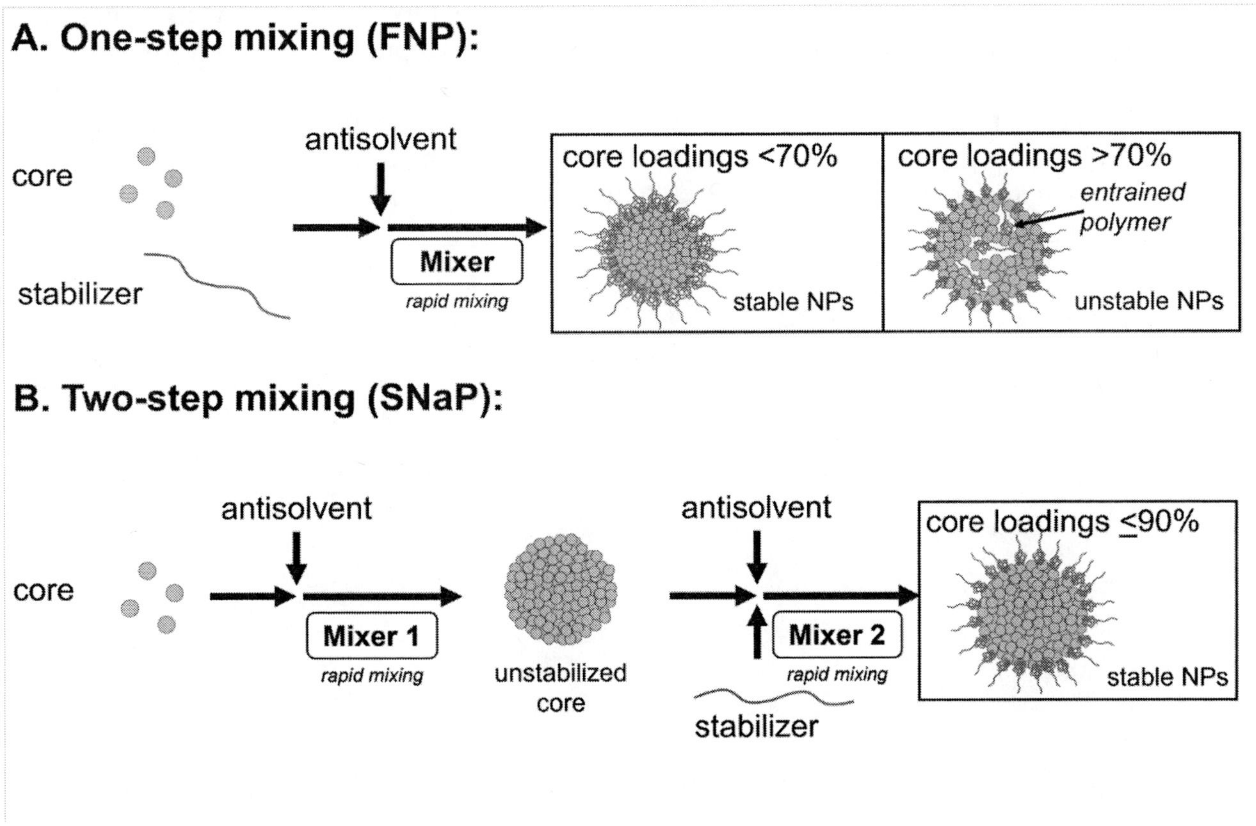

344b Accelerated Development of Continuous Flow Synthesis of APIs with Multi-Objective Bayesian Optimization Using Automated Dynamic Experimentation

Kakasaheb Nandiwale, Cameron Armstrong, Aritra Sarkar, Angel Diaz, Samuel Britner, Federico Florit, Andrew Zahrt, Jason Mustakis, Eric Hansen, Klavs Jensen, Steven Guinness

Continuous flow synthesis of pharmaceuticals is attractive due to its small physical footprint, safety, and consistent product quality with precise control of reaction conditions. The lab-scale development of flow API/intermediate synthesis involves optimization of reaction conditions such as reaction temperature, residence time, and stoichiometry, which are traditionally achieved with multiple experiments analyzed at the steady-state. This can be time-consuming and may consume considerable amounts of materials. Herein we present data-rich experimentation to optimize chemical reactions by means of data-driven experiments. Machine learning (ML) guided Bayesian optimization method is developed for continuous tubular reactors operated under dynamic conditions. We designed in-house LabVIEW™ Virtual Instrument (VI) automation software to control various equipment including Polar Bear Plus Flow™ reactor (Uniqsis), feed pumps, and in-line Agilent UHPLC. In addition, we enabled automated multi-dimensional sinusoidal dynamic experimentation in the LabVIEW™ VI.

We present various case studies of automated multi-objective Bayesian optimization using dynamic experimentation for the continuous flow synthesis of APIs at Pfizer as a part of our Flexible API Supply Technology (FAST) effort. The automated sinusoidal dynamic experiment is performed in PFR to collect data of reaction conditions (e.g., residence time and reagent equivalents). The automated sampling of the design space over a parameters trajectory is achieved with Agilent 1290 Infinity II UHPLC system. We also used flow-NMR to identify impurities and track the product concentration. The information obtained for the dynamic experiment is used for both data-driven optimization and chemical kinetics analysis. We will discuss benefits of Bayesian optimization using automated dynamic experimentation for acceleration of development timelines, while reducing the amounts of raw materials consumed, compared to the steady-state approach.

344c Laser-Driven Innovation: Advancing Pharmaceutical Synthesis with Fiber-Optic Photoreactors

Sergio Carrillo De Hert, Rafael Lopez-Rodriguez, Mike DiMaso, Jonathan McMullen, Steven Ferguson

In the wake of the renewed interest in radical chemistry, photocatalysis has regained attention as a viable method for facilitating environmentally friendly and efficient chemical transformations. The growing enthusiasm for photochemistry arises from its capacity to unlock reactions and engage a diverse array of substrates that typically resist reactivity in conventional chemical environments. Leveraging photochemistry holds promise for streamlining processes, sustaining gentle reaction conditions, sidestepping risky redox agents, and upholding ambient temperature reactions, thus offering a pathway towards cleaner and safer chemical practices.

The revolutionary progress in chromophore complexes that exhibit exceptional light absorption in the visible spectrum has coincided synergistically with the evolution of LEDs. Nevertheless, the utilization of LEDs presents notable hurdles in the design of photoreactors, including issues related to heat generation, light dispersion, safety and varying emissions. In this context, we present and assess a novel laser-driven Continuous Annular Photoreactor (CAP-Flow), which tackles these obstacles by employing radial light diffusing fibres (LDF) coupled with a laser diode through a fibre- optic cable.

Laser technology presents numerous compelling advantages compared to LEDs. Notably, lasers emit highly monochromatic light with narrower spectral bands, which leads to significantly enhanced conversion rates in photochemical reactions and reduces the absorption of photons by unwanted species. Additionally, lasers exhibit low beam divergence, ensuring superior spatial control and minimizing photon loss by avoiding unintended areas of illumination.

Moreover, the operational convenience of transporting photons generated elsewhere into the synthesis area through optic fibres streamlines processes, enhancing overall operational ease. The use of optical fibre-coupled diode lasers offers additional benefits beyond heat management. Firstly, it eliminates the need to position electrical equipment close to the photoreactor, thereby enhancing safety by reducing the risk of ignition near flammable solvents. Secondly, this setup has the potential to obviate the requirement for ATEX solutions in production areas.

The CAP-Flow system is highly modular and capable of continuous operation. It is constructed using standard fluidic components, with the Laser Diode Fiber (LDF) positioned inside and safeguarded by transparent FEP tubing. This tubing traverses through a stainless steel cylindrical housing, forming an annulus. The visual representation of the CAP-Flow system and a schematic depicting its cross-sectional area are illustrated in Figure 1a and Figure 1b, respectively.

The efficacy of the CAP-Flow system's architecture was assessed on a small scale utilizing a 500 mW laser diode, employing ferrioxalate actinometry and the C-N cross-coupling of 4-bromobenzotrifluoride and pyrrolidine as a model application. Actinometry experiments revealed that the CAP-Flow system surpassed other flow technologies reliant on LEDs and lamps. Additionally, the sample reaction demonstrated a Quantum Yield (QY) of up to 3 mol einstein^{-1}, a value only exceeded in the literature by a laser-driven Continuous Stirred-Tank Reactor.

The well-defined geometry and photon irradiation fields within the CAP-Flow system enable mathematical modeling. A partial differential equation (PDE) model was developed, coupling the velocity profile, irradiation profile, and reaction kinetics with the equation of continuity. Solving the PDE provides the ability to track fractional conversion as a function of both radial and axial positions within the reactor, as illustrated in Figure 2(a). Crucial parameters such as kinetic constants and diffusivity can be derived from experimental data using the PDE model and the least squares method. This facilitates the optimization of photocatalyst loading and reactor geometry. Figure 2(b) illustrates iso-conversion curves for different photocatalyst concentrations and mean residence times.

A kg-scale CAP-Flow system has been assembled to evaluate the scalability of the lab-scale prototype. The CAP-Flow's volume was increased sixteen-fold, while the maximum laser power was boosted 250 times, achieved by employing a high-power 500 W blue laser and a distinct type of radial LDF. Figure 3 provides a visual representation of the large-scale CAP-Flow system. The experiments aimed to demonstrate the versatility and scalability of the system, while also validating the predictive accuracy of the PDE model.

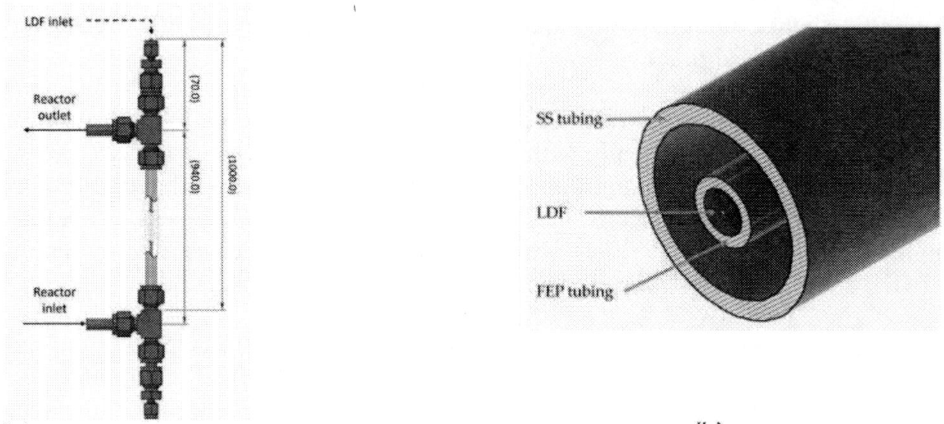

(a) (b)

Figure 1. Architecture of CAP-Flow. (a) Dimensions and (b) cross sectional cut.

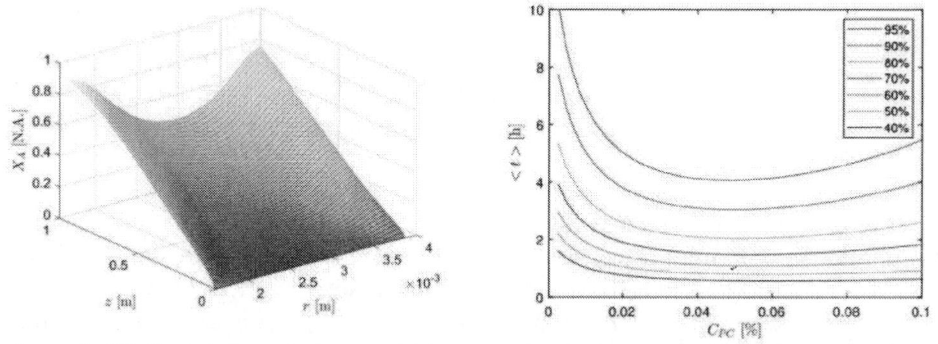

Figure 2. PDE model predictions. (a) Fractional conversion as a function of position and (b) iso-conversion curves for different catalyst loading and mean residence times.

Figure 3. Kg-scale CAP-Flow.

344d Evaluating Pervaporation As an Enabling Technology in Pharmaceutical Processing

Akshay Korde, Christopher Vitale, James Ratway, Jie Chen, Sharad Maheshwari, Manish Kelkar, Eric Moschetta, Nandkishor (Nandu) Nere

Pervaporation is a proven membrane technology that can allow for selective removal of solvent/s from a mixture. Hydrophilic pervaporation membranes have been successfully applied to the dehydration of several solvents, beyond azeotropic compositions. However, pervaporation has seldom been applied to manufacturing processes in the pharmaceutical industry.

Azeo-drying or chase distillation operations for water removal are quite common in the pharmaceutical industry. In this work, we will present the application of pervaporation to replace these distillations to significantly reduce the solvent consumption and subsequent waste generation. The process development in the lab scale and its rational, model-based transfer to pilot scale will be discussed. The impact of different operating modes such as inert gas sweep on the permeate side, pressurizing the feed, and effect of co-solvents to increase the activity co-efficient of water will be discussed. Finally, the challenges associated with implementation of a pilot-scale unit and their potential resolution will be highlighted.

Disclosure:

Akshay Korde, Christopher Vitale, James Ratway, Jie Chen, Sharad Maheshwari, Manish Kelkar, Eric Moschetta, and Nandkishor Nere are employees of AbbVie. All authors may own AbbVie stock. AbbVie sponsored and funded the study; contributed to the design; participated in the collection, analysis, and interpretation of data, and in writing, reviewing, and approval of the final publication.

344e *Automated Liquid Handler Advancements for High-Throughput Scale-Down Perfusion Cell Culture Screening*

Michael Skoumal, Zoe Horton, Orla Gauvain, Elena Shcherbakova, Kyle McHugh, Yang Yang

Perfusion bioreactor technology is highly versatile and can provide increases to productivity in addition to enabling the production of unstable or toxic proteins that are traditionally difficult to produce in fed-batch systems. Process development activities such as process optimization, clone selection, and media screening require a large number of experimental conditions and a representative high-throughput small-scale model is desirable. Automated microscale bioreactor tools like the ambr15 offer a useful option to culture cells, but perfusion cell culture faces challenges due in part to the complexity of mimicking cell separation at larger scales. Progress has been made to solve this issue by utilizing cell sedimentation, which has the benefit of being executed entirely within the ambr15 environment, but it subjects the cells to extended periods of hypoxia due to the long settling time and the relatively slow speed of the single channel ambr15 liquid handler. A faster and more representative alternative is provided by centrifugation of the vessels to pellet the cells and harvest the supernatant with an automated multichannel fluid handler, but this has previouslyrequired significant operator involvement. Here, we developed a workflow consisting of a Tecan Fluent with an integrated centrifuge and custom 3D-printed attachments. Vessels are removed from ambr15 and placed in an automated workstation. Once in the Tecan environment, the entire process of decapping vessels, cell separation by centrifugation, media exchange, and initial media addition is automated and requires no operator interaction. This reduces the time for a vessel to be without agitation, oxygen, and pH control from 90 minutes to 20 minutes. The workflow demonstrated comparable cell growth profiles, productivity, and product quality to our bench-scale intensified N-1 perfusion step. Additionally, the fully automated workflow lowers the barrier to entry for operators by automating a previously manually complex process and has reduced contamination rates from the perfusion step from 10% to 0%.

344f Systematic Characterization of Interfacial Properties of Lipid Nano Carriers of Ternary Lipids (DSPC, Cholesterol and SM-120)

Natsuumi Ito, Kouta Yamaguchi, Nozomi M. Watanabe, Hiroshi Umakoshi

Abstract #692992

Systematic Characterization of Interfacial Properties of Lipid Nano Carriers of Ternary Lipids (DSPC, Cholesterol and SM-120)

Natsuumi Ito[1], Yamaguchi Kouta[1], Nozomi M. Watanabe[2] and Hiroshi Umakoshi[2], (1)Div. of Chem. Eng., Grad. Sch. of Engineering Science, Osaka University, Toyonaka, Osaka, Japan, (2)Div. of Chem. Eng., Grad. Sch. of Eng. Sci., Osaka University, Toyonaka, Osaka, Japan

Abstract Text:
Nanostructured lipid carriers (NLCs) are a new generation of lipid vectors for drug delivery systems (DDSs), which are composed of solid and liquid lipids dispersed throughout the inner lipid matrix. This study provides a systematic method to characterize the interface of the NLC. Twitterionic lipid (DSPC), cholesterol and cationic lipid (SM102) were selected as the target NLC. After the preparation of the NLCs with different compositions, the obtained NLCs were applied to (1) Physical Property (DLS), (2) Physicochemical Properties (Steady-State Fluorescence Analysis: Hydrophobicity and Fluidity with Phase State Estimation based on Previous data), and (3) Surface-Hydrated Properties (Time Resolved Emission Spectra Analysis). A series of data set for various NLCs were shown in different triangle diagrams of the above three components. These systematic results can be used to clarify the design space of the NLC composition to show the appropriate interfacial properties for DDSs.

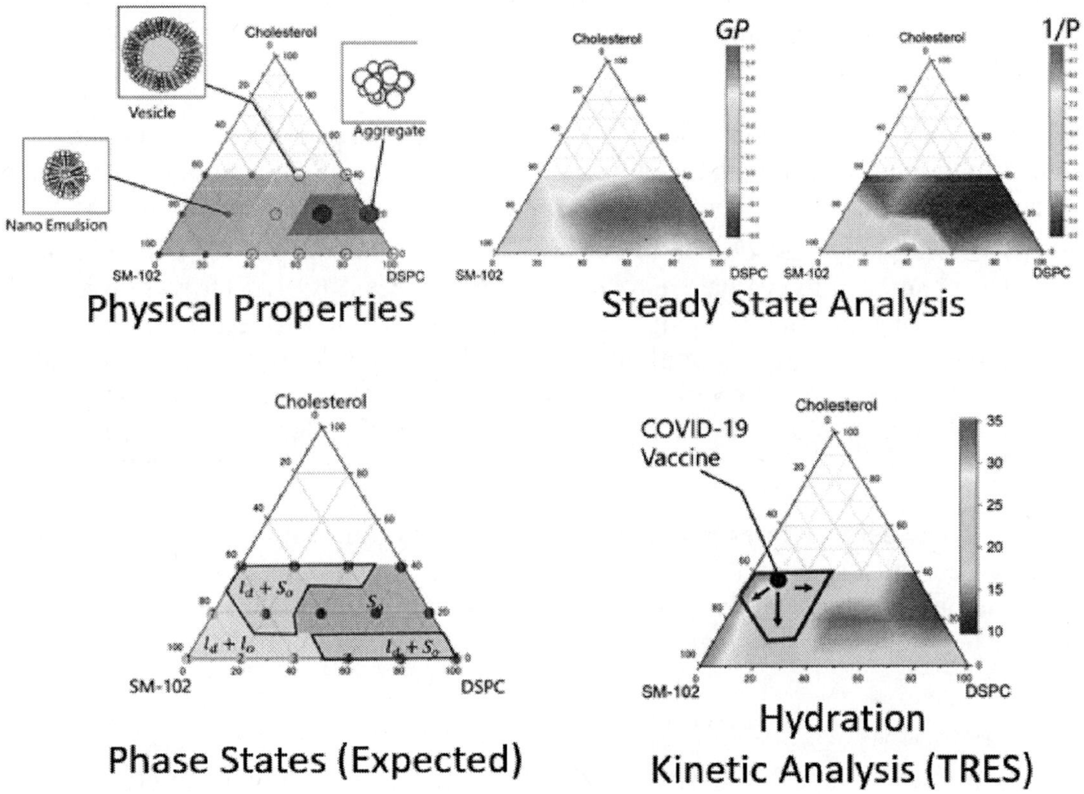

344g Multi-Objective Bayesian Optimisation for Sustainable Reaction Development

Jiyizhe Zhang, Alexei A. Lapkin

Optimising chemical reactions to find the best conditions is important at multiple stages of reaction development, from lab-scale condition screening to large-scale manufacturing. Often multiple objectives need to be considered alongside yield optimisation, for example, achieving high yield with high selectivity, considering economic and sustainability aspects at the same time, or optimizing yield and throughput simultaneously for scale-up. Finding such optimal solutions usually requires searching in a large design space for a set of Pareto solutions rather than one solution, which tends to be tedious and time-consuming. In this study, we focus on optimising two chemical reactions in flow with multiple objectives using Bayesian optimisation, to consider space-time-yield and sustainability factor at the same time.

Amide bond formation is one of the most prevalent reactions in pharmaceutical industry, among which the Schotten-Baumann reaction with a long history is useful as a potential green amide formation approach. However, the use of water in the reaction system often causes undesired hydrolysis and can generate a multiphase system. This makes the reaction space complex and challenging to find the optimal conditions. In the first case study, a Schotten-Baumann reaction was studied in continuous flow and was optimised with two objectives using a Bayesian optimisation algorithm based on the *q*-Noisy Expected Hypervolume Improvement (*q*NEHVI)[1]. The algorithm guided the experiment design over a range of mixed variables, i.e. electrophiles, equivalents, solvents, and flow rates, and was able to identify the Pareto front of optimal solutions efficiently. Based on the optimisation results, reaction under flow and batch conditions were compared; undesired hydrolysis was suppressed successfully using the flow conditions. Finally, the relationship between solvent and flow rate was discussed to gain more insights into this reaction.

In the second case study, we focus on robust applications where large noises are unavoidable. This is useful when dealing with unstable reagents and for large-scale manufacturing processes. To cope with large noises, a multi-objective Bayesian optimisation algorithm based on Euclidian expected quantile improvement (MO-E-EQI[2]) was adopted to consider heteroscedastic noises. First, an *in silico* benchmarking was conducted to evaluate algorithm performances. MO-E-EQI was compared to two state-of-the-art multi-objective Bayesian algorithms and showed the best performance for a noise level up to 20%. Then MO-E-EQI was implemented in a real case to optimize an esterification reaction in flow with four continuous variables: flow rates, equivalents, catalyst loadings, temperatures. The algorithm efficiently identified a clear trade-off under noises between space-time yield and E-factor.

In conclusion, the study demonstrates multi-objective reaction optimisation using Bayesian algorithms for two complex systems: multi-phase reaction system and noisy system. A set of Pareto front solutions can be found efficiently guided by Bayesian approach, which shows a promising way to allow multiple factors to be considered at early-stage reaction development.

Reference

[1] S. Daulton, M. Balandat, and E. Bakshy. Parallel Bayesian optimization of multiple noisy objectives with expected hypervolume improvement. Advances in Neural Information Processing Systems 34, 2021.

[2] D. Semochkina, A. I. Forrester, D. C. Woods. Multi-objective optimisation using expected quantile improvement for decision making in disease outbreaks. arXiv preprint arXiv:2401.12031, 2024.

343a Continuous Flow Synthesis of Grignard Reagents Towards the Preparation of Advanced Intermediates and Active Pharmaceutical Ingredients – a Modular Approach

Thomas Toupy, Darren Whitaker, Mohammed Janoudi, Ted Raymond, Neda Nazemifard, Charles D. Papageorgiou

Grignard reagents remain a cornerstone for carbon-carbon bond formation in organic synthesis and in the pharmaceutical industry; however, their traditional batch preparation suffers greatly from safety concerns due to the pyrophoric nature of magnesium, the exothermic nature of the reaction, and the corrosive nature of the final solution. In recent years, the use of continuous flow chemistry and continuous flow manufacturing strategies have been used to palliate for these limitations. This work explores the application of continuous flow chemistry coupled with process analytical tools (PAT), chemometric and mechanistic modelling for the safe, efficient, and scalable generation of Grignard Reagents and their subsequent use in telescoped processes. Herein, we demonstrate the in-situ activation of magnesium metal and its subsequent reaction with two organic halides R-Br (**1**) and R-Cl (**6**) under continuous flow conditions using PAT (in-line IR and NMR) for reaction monitoring and process control.

Case Study 1 (**Figure 1a**): Upon optimization of the reaction parameters under conventional batch conditions and transpositions under continuous flow, we managed to increase the conversion into our desired Grignard species (**2**) from ca. 85% up to 96%. Then, using tailored chemometric and mechanistic models to monitor the various species involved, the process was run in a closed-loop system under model predictive control for ca. 48h to mimic a manufacturing run. This extended run yielded a daily productivity of up to 1.5 kg of the final material after downstream purification and highlighted the efficiency and robustness of the developed modular platform and process.

Case Study 2 (**Figure 1b**): The modular platform developed in Case Study 1 was repurposed for the preparation of a second Grignard reagent. Upon optimization of the reaction parameters and transposition under continuous flow, the conversion into the corresponding Grignard species (**7**) increased from ca. 85% up to 98%. Following the same strategy of combining chemometric and mechanistic models to monitor the various species, the platform was run for 10h yielding a productivity of up to 710 g a day in the corresponding Grignard species. A 7-day run of the platform is planned to mimic the duration of a manufacturing run. Successful execution will demonstrate the robustness and manufacturing-readiness of the devised modular platform.

This work demonstrates the significant improvement achievable for Grignard reagent generation through the combination of continuous flow chemistry, PAT, and modelling compared to traditional batch methods. In two case studies, the conversion of organic halides to the desired Grignard reagents increased substantially (up to 98% from 85%). Furthermore, continuous flow operations with PAT and model-based control enabled multi-day operation, achieving daily productivities of 1.5 kg and 710 g for the respective Grignard products. These results highlight the efficiency and robustness of the developed platform, paving the way for the safe and scalable manufacturing of Grignard reagents within the pharmaceutical industry and beyond. Future work will focus on further extending the platform's capabilities and demonstrating its robustness through extended manufacturing-mimicking runs for the second case study. This will solidify the approach as a viable and advantageous alternative to traditional batch methods for Grignard reagent generation.

Figure 1a

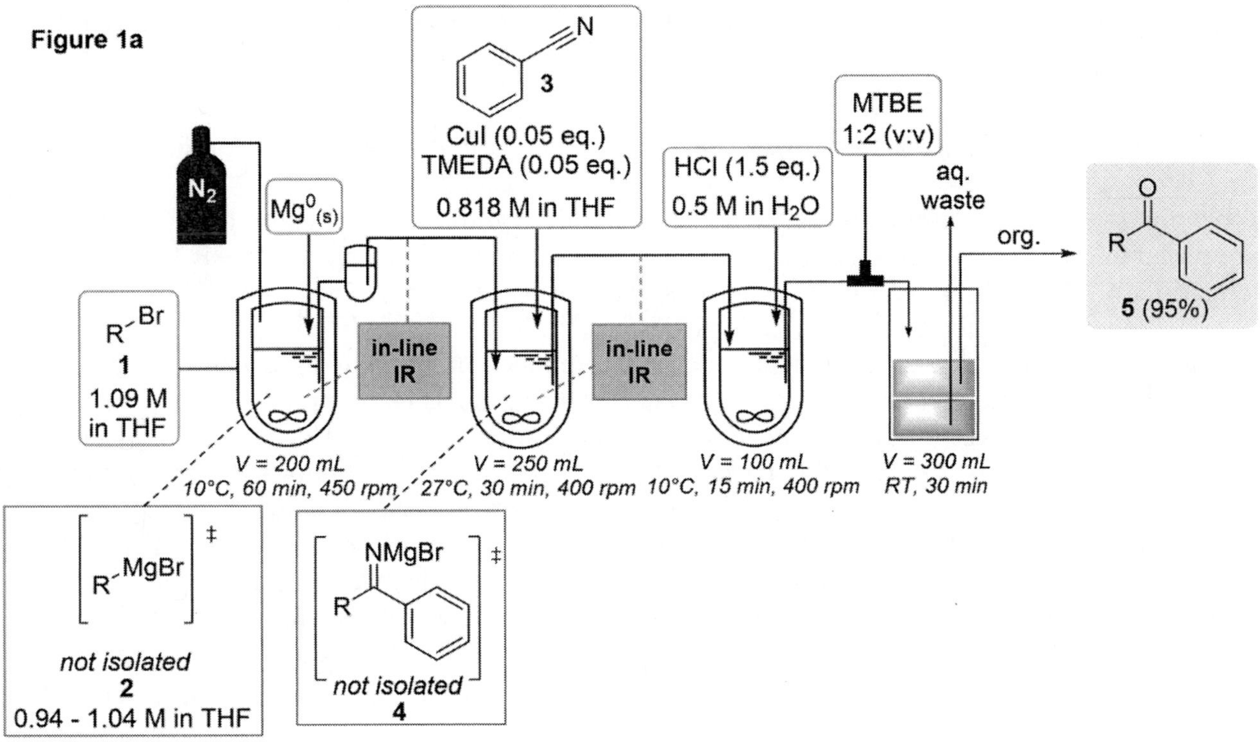

Figure 1b

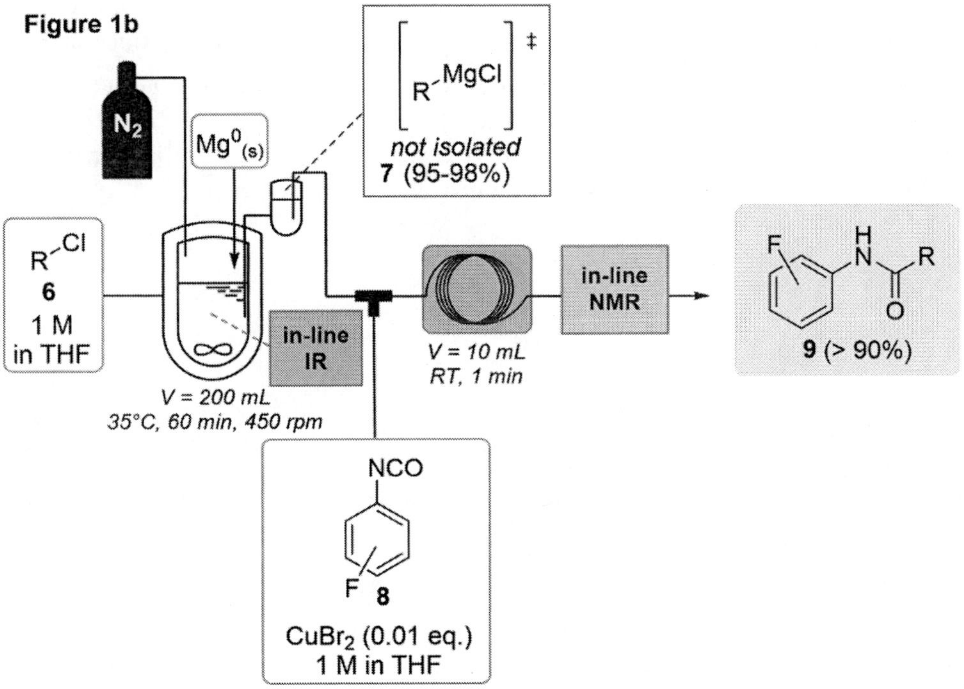

343b API Synthesis in Flow Enabled By on-Line Proton NMR and UV-Vis Chemometrics

Daniel Gregory, Kimberly Penzer, Justin Turnage, Kaitlin Kay, Frank Gupton, James K. Ferri

Active pharmaceutical ingredients (APIs) are traditionally synthesized *via* batch processing strategies as opposed to continuous manufacturing within flow reactors. While batch synthesis facilitates the production of APIs in bulk metric ton quantities, the iterative nature of this methodology often results in extensive hands-on labor, complicated logistics, and vulnerable supply chains. Continuous API manufacturing is emerging as a promising alternative to batch processing as it enables live, on-line monitoring of chemical species from starting materials to products using process analytical technologies (PAT) such as nuclear magnetic resonance (NMR) and UV-Vis spectroscopy. This talk examines the transfer of a traditionally batch API synthesis route to continuous flow manufacturing for the production of albuterol sulfate within a modular continuous API manufacturing system. Albuterol is a beta-agonist bronchodilator used for the treatment of asthma and frequently listed on the ASPR drug shortage list. Practical aspects related to appropriate synthon selection to enable flow chemistry, process intensification, and on-line analytical characterization will be discussed. Specifically, we will highlight the utilization of ^{1}H NMR and UV-Vis as powerful on-line PAT tools to assess S_N2 amination chemistry, catalytic hydrogenation, and purity assessment of the final API product in flow. Methods for improving ^{1}H NMR data quality and resolution *via* carbon satellite decoupling and solvent suppression will be demonstrated. Additionally, an overview of signal processing and chemometric modeling is used to establish a quantitative correlation between high performance liquid chromatography (HPLC) with both ^{1}H NMR and UV-Vis for on-line concentration monitoring (*i.e.*, 1.0-50 mg/mL). Lastly, practical engineering design considerations related to the closed-loop integration of these systems into a continuous end-to-end manufacturing system will be discussed to help mitigate shortages of this critical lifesaving drug.

343c Effects of Tracer Physical Properties and Quantities on the Residence Time Distribution (RTD) of a Continuous Pharmaceutical Manufacturing Line

Yi Tao, Sonia M. Razavi, Carlos Ortega-Zuniga, James Scicolone, Fernando Muzzio

Over the last decade, the pharmaceutical industry, along with its technology, ingredient, and equipment suppliers, has embraced a worldwide transformation from batch to continuous manufacturing (CM). This shift is due to its potential for enhancing agility, flexibility, and robustness in both product and process development [1-3]. Regulatory agencies, such as the Food and Drug Administration (FDA), have also supported this transition. Starting in 2004, the FDA progressively endorsed CM, from revising the Process Analytical Technology (PAT) guidance to comprehensive guidelines in 2021 [4]. Recent FDA data revealed that drug approvals for products made via CM are, on average, eight months faster than those produced by traditional batch processes [5].

Residence Time Distribution (RTD) is a tool used to understand process dynamics, including understanding of propagation of disturbances through each unit and the degree of back-mixing in a continuous flow system [6]. This is significant for maintaining a state of control [7]. Characterizing RTDs can be utilized to develop control strategies, implementing PAT to track disturbances to products and divert out-of-specification (OOS) materials [6, 8]. A crucial part of RTD characterization involves selecting a tracer representative of an active pharmaceutical ingredient (API) and injecting the tracer into the system without interfering with the bulk flow.

This work continues previous efforts on tracer selection for continuous processing [9, 10], which focused solely on RTD characterization within a continuous blender. Expanding upon this preliminary work, our study focuses on investigating the effects of both physical properties and quantities of the tracer on the RTD of a continuous direct compression (cDC) manufacturing line. An easy-flowing formulation was used: 50% w/w Compap L™, 44% w/w Prosolv® HD90, 5% w/w pregelatinized starch, and 1% w/w MgSt. Three tracers were selected from a large material property database by quantifying material similarity between the API (i.e., Compap L) and other materials (i.e., tracer candidates), two of which were most similar to the API, and the third was dissimilar. For each tracer, 6, 15, and 30g were used. Each quantity was individually injected into a blender inlet, and the concentration of the tracer was measured in the feed frame of a tablet press. RTD metrics, including Mean Residence Time (MRT) and time delay, were analyzed to determine the statistically significant effect of tracer quantities and properties using Analysis of Variance (ANOVA). The complete RTD profiles were compared across different tracer masses and materials using Multivariate Analysis of Variance (MANOVA). The results demonstrated that the amount of tracers and tracer properties did not significantly impact the RTDs measured in the feed frame. This can be explained by the feed frame mixing capability, which effectively disperses the disruption caused by the tracer pulse, due to the prolonged residence time of tracers inside the feed frame. The significance of this work is to expand the methodology for choosing the right tracers for the RTD of cDC: for an easy-flowing formulation, we can select a broader range of materials as tracers and inject larger amounts of tracers with better reproducibility without challenging PAT detection limits.

Reference:

1. Lee, S.L., et al., Modernizing pharmaceutical manufacturing: from batch to continuous production. Journal of Pharmaceutical Innovation, 2015. 10: p. 191-199.
2. Ierapetritou, M., F. Muzzio, and G. Reklaitis, Perspectives on the continuous manufacturing of powder-based pharmaceutical processes. 2016, Wiley Online Library.

3. Vanhoorne, V. and C. Vervaet, Recent progress in continuous manufacturing of oral solid dosage forms. International Journal of Pharmaceutics, 2020. 579: p. 119194.
4. Muzzio, F. and S. Oka, How to design and implement powder-to-tablet continuous manufacturing systems. 2022: Academic Press.
5. Fisher, A.C., et al., An audit of pharmaceutical continuous manufacturing regulatory submissions and outcomes in the US. International Journal of Pharmaceutics, 2022. 622: p. 121778.
6. Gao, Y., F.J. Muzzio, and M.G. Ierapetritou, A review of the Residence Time Distribution (RTD) applications in solid unit operations. Powder technology, 2012. 228: p. 416-423.
7. GUIDELINE, I.H., Continuous Manufacturing of Drug Substances and Drug Products Q13. 2021, ICH.
8. Engisch, W. and F. Muzzio, Using Residence Time Distributions (RTDs) to Address the Traceability of Raw Materials in Continuous Pharmaceutical Manufacturing. J Pharm Innov, 2016. 11: p. 64-81.
9. Razavi, S.M., et al., Selection of an appropriate tracer to measure the residence time distribution (RTD) of continuous powder blending operations. Powder Technology, 2023. 429: p. 118864.
10. Escotet-Espinoza, M.S., et al., Effect of tracer material properties on the residence time distribution (RTD) of continuous powder blending operations. Part I of II: Experimental evaluation. Powder technology, 2019. 342: p. 744-763.

343d A Universal Control Concept for Complex Manufacturing Lines

Stephan Sacher, Lisa Kuchler, Atabak Azimi, Hazem Damiri, Martin Steinberger, Martin Kureck, Dirk Kirschneck, Johannes Khinast, Andreas Kottlan

With the implementation of advanced manufacturing concepts such as continuous operation, integration of different process routes or modular manufacturing, the control of the process and the related product quality becomes more complex. In modern high-speed manufacturing, no delays from measurements can be afforded between unit operations as with traditional in-process controls or end-product testing. Real-time information on the material state and associated control actions are required.

This talk will present a universally applicable control concept, which allows the highest level of flexibility for processing, equipment and automation level. The concept can deal with a combination of batch-wise and continuously operated process steps, with manual and automated operation, with different scales of equipment and with digital and manual data acquisition. It is demonstrated at a compact end-to-end manufacturing line for tablets. The control strategy is not only based on data from classical PAT sensors, but also from equipment and from soft sensors. As an example, the content is monitored and controlled using soft sensors and the feeder data. A digital twin of the entire process chain is fed with process and quality information and executes the control actions. The complete process control is realized in an automation platform with human machine interface (HMI). Additionally, some novel process routes have been realized: A suspension of active pharmaceutical ingredient (API) is directly fed into hot melt extrusion (HME), which is combined with direct compaction (DC).

343e Wavelength Selection Algorithm Robust Against Impurities for Sample-Efficient PAT

Sakuya Kobayashi, Shota Kato, Manabu Kano

Introduction

Near-infrared (NIR) spectroscopy is a method that can analyze samples faster than conventional methods such as high-performance liquid chromatography (HPLC) and gas chromatography (GC). Building calibration models requires a lot of materials and time. To solve this problem, there has been a growing interest in pure component-based approaches such as classical least squares (CLS) [1] and iterative optimization technology (IOT) [2] due to their cost-effectiveness in the pharmaceutical industry. However, these approaches may not work when impurities exist because they assume no impurities. In the present research, we propose a new wavelength selection method that makes pure component-based approaches robust against impurities with as few samples as possible.

Conventional pure component-based method

Both CLS and IOT are pure component-based modeling approaches.

CLS is a regression-based method based on the Beers-Lambert law. CLS regression coefficients are determined by using pure spectra. However, when CLS predicts concentrations of components, there is a possibility of generating unrealistic results such as negative concentrations.

IOT was developed to solve the problem of CLS. Instead of using regression modeling, IOT solves an optimization problem to make predictions under constraints, preventing unrealistic values from being generated. Typical constraints include:
C1. The mole fraction of any pure component is greater than or equal to 0 and less than or equal to 1.
C2. The sum of the molar fractions of known pure components is equal to 1.

Proposed method

IOT assumes that impurities are not included. When impurities are generated by side reactions, however, the sum of the mole fractions of known pure components will be less than 1. This situation violates constraint C2. To solve this problem and make IOT robust against impurities, we propose to use IOT without constraint C2.

In addition, to predict molar fractions accurately even if impurities are present, we propose a new wavelength selection method with a few samples, which is referred to as sample-efficient wavelength selection (SEWS). The samples are defined as sets of spectra of mixtures and concentrations of the known pure components. The procedure of SEWS is as follows.

1. Remove the influence of known pure components from the spectra of mixtures. The mixtures consist of known pure components and impurities. The spectra of the known pure components are given, those of the impurities are unknown. The number of mixture spectra is N, which is the number of samples. The remaining spectra are denoted as A_{other}.

2. Calculate the variance of A_{other} at each wavelength.

3. Select the wavelengths where the variance of A_{other} does not exceed a certain threshold. The other wavelengths are considered unsuitable for a pure component-based approach and is not utilized in model construction.

In this work, we combine IOT without constraint C2 with SEWS to predict the mole fractions accurately in the presence of impurities. This method is called SEWS-IOT.

Results and Discussion

To verify the performance of SEWS-IOT, simulated data that meet the following conditions were generated.

1. There are three components: materials A, B, and C. The pure spectra of materials A and B are known, but the pure spectram of material C is unknown.

2. The peaks of the pure spectra of materials A and B partially overlap with the peaks of the pure spectram of material C.

3. The spectrum of a mixture follows the Beer-Lambert law.

The sample size N for wavelength selection was 0, 5, or 10, and that for testing the prediction model was 100. We used SEWS-IOT to predict the molar fraction of material A. This experiment was repeated 10 times by changing datasets (random seeds), and the average RMSE was calculated for the performance evaluation.

When SEWS was not used (i.e., $N = 0$, conventional IOT), the average RMSE value was 3.0×10^{-2}. As the number of samples N increased, the average RMSE decreased. The average RMSE was 2.1×10^{-2} when $N = 5$, and it was 1.5×10^{-2} when $N = 10$. By using 10 samples, SEWS-IOT reduced the average RMSE by 50% in comparison with IOT without constraint C2. The results have demonstrated the effectiveness of SEWS as a sample-efficient wavelength selection method.

Conclusion

We proposed SEWS, a sample-efficient wavelength selection method robust against impurities, and we showed the effectiveness of SEWS-IOT, which can enhance the prediction accuracy of IOT using SEWS. The proposed SEWS-IOT reduced the average RMSE by 50% with just 10 samples for wavelength selection. By using this proposed method, process analytical technology (PAT) can be realized economically when conventional pure component-based methods do not work due to impurities. In addition, spectroscopic analysis such as NIR can be easily applied to analyzing the results of chemical reactions where impurities are generated.

Acknowledgement

This work was supported by New Energy and Industrial Technology Development Organization (NEDO) Grant No. JPNP19004.

References

[1] Adam, J., et al.: Comparison Between Pure Component Modeling Approaches for Monitoring Pharmaceutical Powder Blends with Near-Infrared Spectroscopy in Continuous, AAAP Jornal, 24 (2022)

[2] Muteki, K., et al: Using Iterative Optimization Technology (Calibration-Free/Minimum Approach, Industrial and Engineering Chemistry Research, 52, 12258–12268 (2013)

343f Advanced Manufacturing: Mechanistic Understanding and Optimization of a Twin-Screw Wet Granulation Process for Manufacturing of Extended-Release Tablets

Nobel Sierra Vega, Muhammad Ashraf, Thomas O'Connor, Ahmed Zidan

Advanced manufacturing has the potential to revolutionize the pharmaceutical industry by enhancing drug quality, addressing shortages of medicines, and accelerating time-to-market. The U.S Food and Drug Administration (FDA) actively encourages the adoption of advanced technologies to ensure the safety and efficacy of drug products. Continuous manufacturing has emerged as a focal point due to its ability to seamless process scale-up, increase the production efficiency, and improve the process control. However, its successful implementation requires a comprehensive understanding of the process dynamic. Twin-screw wet granulation (TSWG) is a continuous manufacturing technology that may be implemented in a fully continuous from-powder-to-tablet production line. While successfully used in the commercial manufacture of immediate-released tablets; implementing TSWG for extended-release (ER) drug products has presented challenges due to the limited options for setting the design and operating parameters to processing the ER formulations. This presentation delves into a systematic approach for defining an optimal design space for TSWG processes, focusing on the interplay between design and operating parameters and their impact on process dynamic and critical characteristics of granules and ER tablets.

A case study involving the processing of an ER formulation with high load of swellable polymers is presented. Through a definitive screening design and principal component analysis, the influence of screw speed, powder feeding rate, and the number of kneading (KEs) and sizing elements on residence time distribution (RTD) parameters and granules and tablets characteristics was investigated. The results demonstrated that screw design and operating parameters affected the flow and bulk characteristics of granules, while the screw speed and KEs were the most significant factors affecting the RTD parameters. Dissolution profiles revealed that granule characteristics mainly influenced the early-phase drug release, while formulation matrix dominated the extended drug release phase. In summary, this presentation not only underscores the importance of simultaneous optimization of operating and design parameters for the manufacturing of ER tablets with desired performance characteristics, but also emphasizes in the notable failure modes in this process. Furthermore, strategies to incorporate quality-by-design principles to monitor and control the critical quality attributes of granules are discussed.

DISCLAIMER

This presentation reflects the views of the presenter and should not be construed to represent FDA's views or policies.

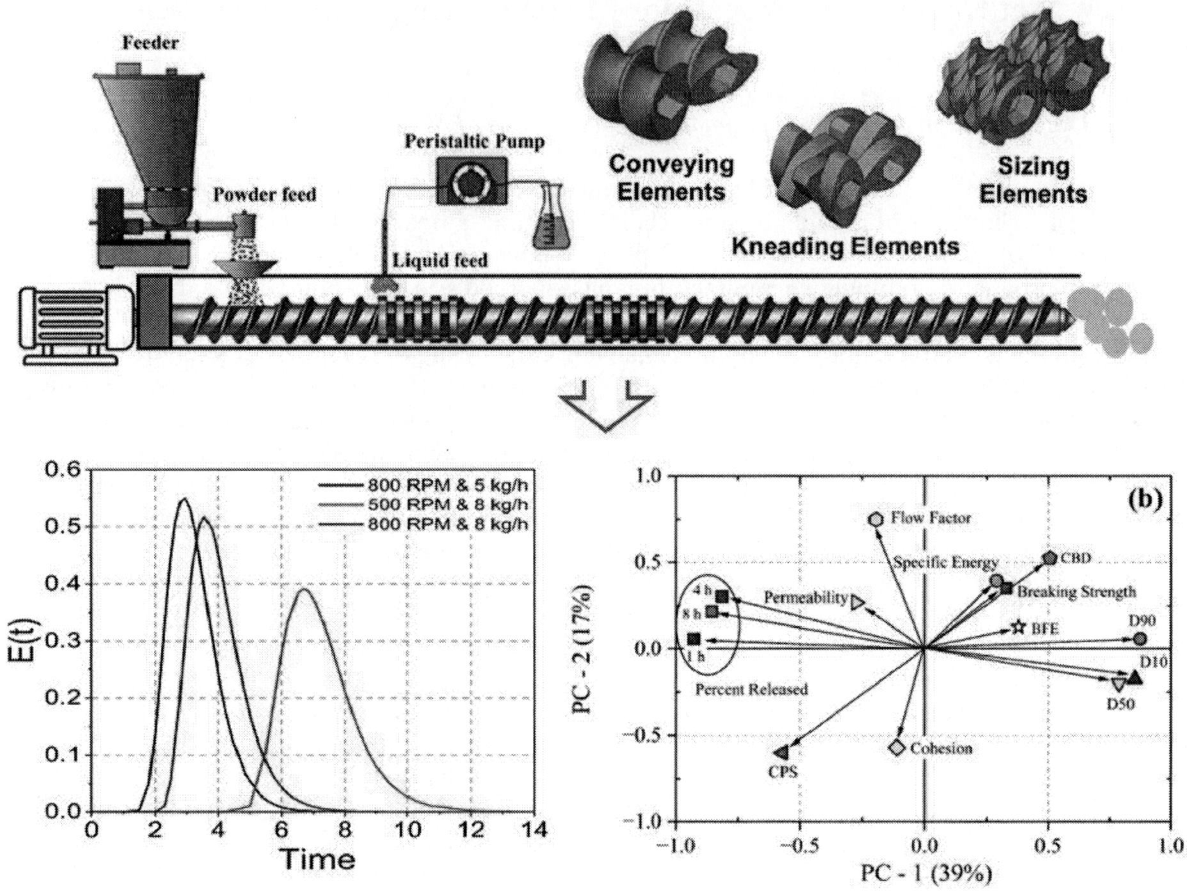

Feeder
Powder feed
Peristaltic Pump
Liquid feed
Conveying Elements
Kneading Elements
Sizing Elements
800 RPM & 5 kg/h
500 RPM & 8 kg/h
800 RPM & 8 kg/h
E(t)
Time
(b)
Flow Factor
Specific Energy
CBD
Breaking Strength
Permeability
BFE
D90
D10
D50
4 h
8 h
1 h
Percent Released
Cohesion
CPS
PC - 2 (17%)
PC - 1 (39%)

343g Application of Fluidized Bed Drying Modeling to the Development and Scale-up of a Continuous Wet Granulation Line

Marta Moreno Benito, Peiyuan Liu, Mark Pinto, Jenna Pasko, Kai Lee, Daniel O. Blackwood, Pankaj Doshi

Continuous wet granulation has gained attention in the last few years for the continuous manufacturing of high drug load products. Continuous manufacturing is associated with a reduction of time and cost for drug development and manufacture using more sustainable processes, with minimal scale-up and technology transfer [1] and hence has made a significant progress and won researchers' attention [2] in the last decade. The simplest set up is direct compression route, consisting of powder feeding, blending, and tableting. In the case of wet granulation, this is extended by incorporating intermediate steps such as wet granulation, drying, and milling. This improves the material flow, compaction, stability, and reduces sticking, among others, which is especially relevant in high drug load blends. To maintain reduced development times, process modelling is key to identify, reduce, and control risk in the different process stages. A part of the process where modelling can significantly contribute is the drying of the wet granules. In the continuous wet granulation technology ConsiGma-25 [3], which is part of the Pfizer Continuous Manufacturing Module (PCMM) [4], the drying process takes place in a segmented fluidized bed dryer (FBD). Particularly, a twin-screw system with liquid injection generates the wet granules, which are distributed sequentially in six cells that dry and discharge a specific load of material, targeting a final granule moisture. The resulting granules are then conditioned and fed at constant throughput to the granule blending and tablet compression.

There are several aspects of the drying stage design and de-risking that can be addressed with digital tools, including: (i) the schedule of the segmented six-cell dryer, (ii) potential drying constraints for high throughputs due to the semi-continuous nature of the drying operation, (iii) defining the discharge condition that guarantees the target loss on drying (LOD) is reached in the final dried granules for new formulations, e.g. typically using the so-called delta-T approach, (iv) the estimation of drying times depending on multiple factors like initial granule moisture, drying air conditions, cell load, other material properties, and their potential variability in the process, (v) the tracking of moisture content in downstream granule feeding, mixing and tableting operations, and (vi) understanding the effect of moisture content in the extra-granule mixing behavior.

In this work, several process modelling tools have been applied in the development of the drying process in a continuous wet granulation line at clinical and commercial scale. This includes the scale-up from a lab-scale process to one-cell fluidized bed drying in a simpler ConsiGma-1 system, and full-scale segmented drying in ConsiGma-25, together with the technology transfer from development to manufacturing. Additionally, digital tools are used to understand the impact of cell-to-cell variability, reduce the number of required experiments for new drugs, and improve the process design for continuous drying. Process simulations allow making better predictions, optimize the drying process, and reduce the development time of new compounds.

This way, a first principles FBD model has been applied, which captures the most relevant experimental conditions, like the liquid/solid ratio, cell filling time, and drying air flowrate, temperature, and relative humidity. The model predicts the drying time required to achieve granules with moisture content below a given target consistently in all cell loads. It is based on material and energy balances, related through the drying rate. The model accounts for two differentiated phases, where the free-bound and the close-bound moisture is evaporated respectively, referred to as the falling rate kinetics [5]. It was developed using APAP ConsiGma-25 data with different initial moisture content. Drying parameters and material properties were measured using DVS (Dynamic Vapor Sorption) and MAL (Material Assessment Lab) analysis. Material transference parameters were estimated using measured dryer cell temperature and moisture content from experiments. Experimental data was used for model validation. This model allows us to predict the drying curve and drying time for reaching a target moisture content for different initial granule water contents (Figure 1). Additionally, the model is directly applied to optimize the drying process using virtual Design of Experiments (DoE) and to evaluate the impact of the variability in different material and processing conditions, including granule moisture, particle size, drying air temperature, flow rate and humidity, and cell load, using Global Systems Analysis. Residence time distribution approaches have been used for material tracking. The six-cell scheduling at different throughputs and liquid/solid ratios, as well as the minimum fluidization condition [6] based on Ergun equation are also analyzed.

[1] K. Plumb (2005). Continuous Processing in the Pharmaceutical Industry: Changing the Mind Set. Chem. Eng. Res. Des. 83, 730–738. https://doi.org/10.1205/cherd.04359.

[2] D. Blackwood, (2018). A Check-in on our Journey Towards Creating a Continuous Direct Compression Platform Technology. IFPAC Annual Meeting, 2018, Maryland, March 11-14.

[3] ConsiGma-25 reference

[4] R. Steiner, M. Jornitz (2017). Continuous processing in the pharmaceutical industry: status and perspective. In: Kleinebudde, P., Khinast, J., Rantanen, J. (Eds.), Continuous Manufacturing of Pharmaceuticals. John Wiley & Sons Inc, pp. 369–403. https://doi.org/10.1002/9781119001348.ch11.

[5] J. Burgschweiger, E. Tsotsas (2002). Experimental investigation and modelling of continuous fluidized bed drying under steady-state and dynamic conditions, Chemical Engineering Science 57 (24), 5021-5038.

[6] S.U. Pradhan, J.W. Bullard, S. Dale, P. Ojakovo, A. Bonnassieux (2022). A scaled down method for identifying the optimum range of L/S ratio in twin screw wet granulation using a regime map approach, International Journal of Pharmaceutics 616, 121542

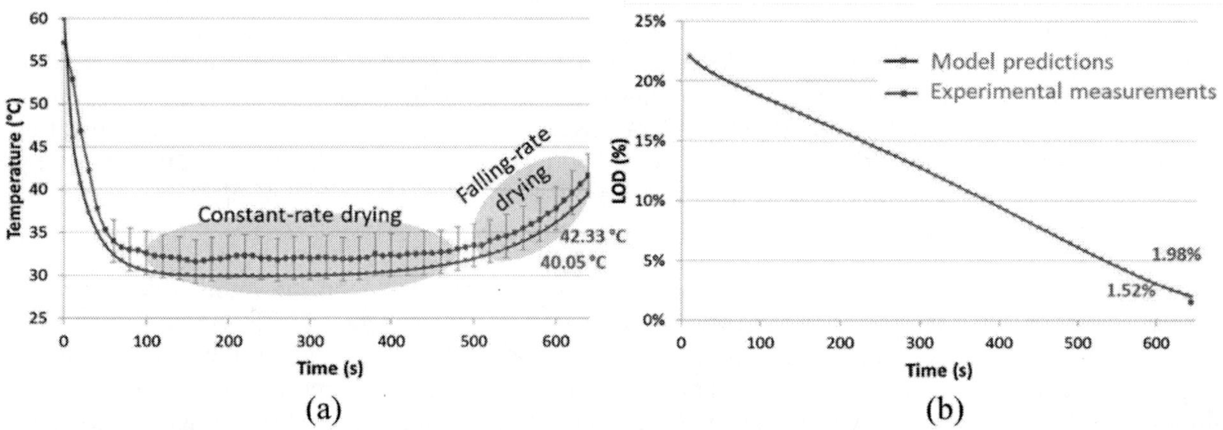

Figure 1: Comparison of simulated and experimental data of: (a) dryer cell temperature and (b) granulate moisture content.

343h Application Design of Experiments Methodologies in Pharmaceutical Development, Manufacturing, and Regulatory Assessment: Current Status, Opportunities and Challenges

Huiquan Wu, Daniel Obrzut

Design of Experiments (DOE) is a well-established yet essential tool for Quality-by-Design (QbD) and Quality Engineering. It has been used across many industrial sectors for robust process and product design and optimization. In the past two decades, DOE has gained popularity during pharmaceutical development and manufacturing to expedite process and product development, ensure process robustness, and facilitate regulatory assessment at both pre-approval stage and post-approval stage. In particular, DOE is beneficial for complicated and advanced manufacturing processes. In this presentation, we will summarize current status, discuss opportunities and future challenges. In addition, we will share several case studies to illustrate some of frequently encountered issues that we have identified through pharmaceutical manufacturing process assessment in the past few years. This work will highlight the benefits of using DOE to develop and manufacture safe, effective, and high-quality medicines.

654a A Graph-Theoretic Approach for Generalized Analytical Solutions of Population Balance Equations with Complex Breakage Kernels in Particle Milling

Prem Podupu, Vamsi Vikram Gande, Ragavendra Hari, Akshay Korde, Sharad Maheshwari, Nandkishor (Nandu) Nere, Meenesh R Singh

Particle size reduction through milling is pivotal in diverse industries like mineral processing, pharmaceuticals, and agrochemicals. Understanding how particle size distribution changes during milling often relies on population balance equations (PBEs) with mechanistic breakage kernels. While numerical methods are common for solving these equations, analytical solutions are usually limited to PBEs with linear breakage kernels. In this study, we introduce a new graph-theoretic approach to develop a generalized analytical solution for PBEs with complex breakage kernels. By breaking down the PBE solution and using Sylvester's formula, we identify the eigenvalues and eigenvectors of the milling matrix based on graph theory principles. The transition rates of particles between sizes, as shown by the elements of eigenvectors, capture the intricate nature of particle breakage, including multiple size changes. This method marks a comprehensive attempt to devise a generalized analytical solution capable of handling all types of breakage kernels, regardless of their linearity. To confirm our method's effectiveness, we used the particle size distribution (PSD) of small molecule crystals from wet-milling to predict the breakage kernels. Further, these breakage kernels were used to validate the PSD at different conditions. We also compared our graph-theoretic solution with a recognized analytical solution for a linear fracture kernel, observing a maximum error of around 3% for a 100-point discretization over extensive time scales. Our study also outlines a broader framework for potential future research. These generalized solutions offer crucial mechanistic insights into milling processes and facilitate the estimation of milling matrix or fracture kernels from experimental data. Importantly, our method is computationally efficient, requiring fewer than 100 lines of code, and can be automated for exploring breakage-related phenomena across various fields.

654b In-Depth Understanding of the Impact of Material Properties on the Performance of Jet Milling of Active Pharmaceutical Ingredients

Kensaku Matsunami, Viktor Bultereys, Laure Descamps, Ashish Kumar

Milling of active pharmaceutical ingredients (APIs) is a critical unit operation to enhance bioavailability, especially of poorly soluble drug substances. Among possible options for milling, air jet milling has the advantages of self-classifying, non-contaminating, and non-degrading operation with a narrow particle size distribution [1]. Within the framework of Quality-by-Design, both experimental and modeling work have been performed to understand the impact of process settings, e.g., mass and gas flow rates, on the milling performance. So far, different types of mechanistic models have been proposed, e.g., population balance modeling (PBM) [1], computational fluid dynamics and discrete element methods (CFD-DEM) [2], and energy-based modeling [3]. While energy-based modeling is much easier and computationally less demanding than other mechanistic models, it cannot consider the impact of material properties on the performance [4]. An in-depth understanding of the impact of material properties on performance is quite relevant since possible experiment numbers using pure APIs are very limited in the design phase.

This study presents the analytical results of spiral jet milling performance varying APIs and process settings. Before the experiments, the material characterization of each API was performed in terms of mechanical properties, e.g., Young's modulus and Poisson's ratio, and energy parameters, e.g., elastic recovery and specific work of compaction. Four different APIs with multiple grades were milled by a Hosokawa Alpine spiral jet mill 50AS. For each API, full-factorial experimental designs were performed changing the mass feed rate and gas feed rate as the factors. Particle size distributions were then characterized by laser diffraction measurements, where the 10th (dv10), 50th (dv50), and 90th (dv90) percentile of the cumulative volume distribution were computed. As indicators of the milling performance, the ratios of dv10, dv50, and dv90 for unmilled APIs to those for milled APIs were calculated.

The experimental results showed that Young's modulus influenced the milling performance. The ratios of dv10 for unmilled to milled APIs became larger, i.e., higher performance when Young's modulus of APIs was higher. While the preferability of less elastic materials could explain this observation, it was also observed that there was collinearity between Young's modulus and the initial particle size in the training data. Regarding the impact of process settings, multi-linear regression models showed that gas flow rate had a higher influence on the milling performance than mass feed rate. The effects of material properties and process settings could be generalized through advanced analysis, e.g., partial least squares (PLS) regression and the PBM calibration.

In conclusion, an in-depth understanding of spiral jet milling was possible through extensive experiments and model-based analysis. The findings can be utilized to design the spiral jet milling process for new APIs with less experimental effort. More detailed and robust insights can be obtained through further detailed statistical analysis, as well as by linking experimental results with the PBM.

[1] S.S. Bhonsale, Bard Stokbroekx, Jan Van Impe. Assessment of the parameter identifiability of population balance models for air jet mills, Computers & Chemical Engineering, 143 (5), 107056 (2020).
[2] S. Bnà, R. Ponzini, M. Cestari, C. Cavazzoni, C. Cottini, A. Benassi. Investigation of particle dynamics and classification mechanism in a spiral jet mill through computational fluid dynamics and

discrete element methods, Powder Technology, 364, 746-773 (2020).
[3] R. MacDonald, D. Rowe, E. Martin, L. Gorringe. The spiral jet mill cut size equation, Powder Technology, 299, 26-40 (2016).
[4] S. Naik, B. Chaudhuri. Quantifying dry milling in pharmaceutical processing: a review on experimental and modeling approaches, Journal of Pharmaceutical Science, 104 (8), 2401-2413 (2015).

654c Computational Strategies in Pharmaceutical Materials Sciences to Derisk Solid Form

Shubham Sharma

A plethora of organic molecules exhibit polymorphism, which refers to the ability of chemical compounds to pack in to different 3D crystalline motifs. This phenomenon is of special importance to both industry and academia since the physical & chemical properties, such a stability, bioavailability & manufacturability are largely related to the crystalline motifs and may vary tremendously between polymorphs. Hence, it becomes crucial for the pharmaceutical industry to understand and flag the risk of finding another polymorph associated to guide the design of a particular medicine. With the advancements in employing predictive sciences for derisking & design of solid forms, several state-of-the-art computational tools have emerged in the past decades. In particular, energetics-based (QM/MD) crystal structure prediction (CSP) and informatics-based health checks have been quite popular across all pharmaceutical research.

CSP has emerged has a pivotal resource within Pfizer's de-risking workflow enabling us to determine the most stable and pharmaceutically desirable crystalline motifs using advanced atomistic modeling approaches. Informatics based health checks, on the other hand, leverages information available through the vast crystallographic structural database available through Cambridge Structural Database (CSD) and Pfizer's inhouse database to flag potential risks in terms of intra- or inter-molecular geometry, hydrogen bonding network, crystal packing, etc. The combination of both the approaches are embedded within Pfizer's derisking workflow and employed to guide design studies.

In this presentation, using case-studies, consisting of both computational & experimental outcomes, we highlight the application of tools like CSP and health checks for designing & derisking of the nominated solid form.

654e Machine Learning (ML) Model for Nozzle Droplet Prediction in Spray Drying

Ankur Kapil, Subhodh Karkala, Michael Hamlin

Approximately 40% of the New Chemical Entities (NCEs) in Pharma are poorly soluble which leads to inadequate bioavailablity[1,2]. There are multiple options to resolve this issue including Hot Melt extrusion, Spray Drying, Cocrystal, Milling, etc. Spray drying is one of the very promising technologies for particle engineering of Pharma API to overcome this solubility issue. Spray drying can be used to generate very fine powders for inhalable drugs and to improve the bioavailability of solid dosage drugs using amorphous solid dispersion. However, spray drying is quite a challenging process to optimize and scale up. The droplet size generated from the nozzle has a significant impact on the final product yield in the spray dryer.

The droplet size from a Spray Drying nozzle depends on the liquid flow rate, nozzle tip size, air flow rate, & solvent. It is not feasible to determine the best operating conditions for any new API with experiments on Spray Dryer for each combination. Many correlations are available to predict the droplet size based on the operating conditions for Spray Drying nozzles[3]. However, each of these correlations results in vastly different results on droplet size distribution and is only suitable for the specific case. Therefore, this work measured the droplet size for different operating conditions and the nozzle specification using laser diffraction. Droplet size was measured for acetone, water, & Amino Methacrylate Copolymer in acetone. A total of 11,390 runs were conducted to evaluate the impact of parameters on the droplet size. In this work Machine Learning (ML) models have been generated to predict the droplet size based on operating conditions and the nozzle design. Simple regression tree, gradient boosted regression tree, Random Forest, & Tree Ensemble were evaluated in this work. The ML model provides a significant competitive advantage to Veranova by significantly reducing the number of droplet size trials required for a new customer product. The best ML models predicted the validation data with 99.4% accuracy for acetone and 92.9% for water solvent respectively.

References:

1. Price, G. & Patel, D. A. Drug Bioavailability. in *StatPearls* (StatPearls Publishing, Treasure Island (FL), 2022).
2. Savjani, K. T., Gajjar, A. K. & Savjani, J. K. Drug Solubility: Importance and Enhancement Techniques. *ISRN Pharm.* **2012**, 195727 (2012).
3. Hede, P. D., Bach, P. & Jensen, A. D. Two-fluid spray atomisation and pneumatic nozzles for fluid bed coating/agglomeration purposes: A review. *Chem. Eng. Sci.* **63**, 3821–3842 (2008).

654f A Holistic Predictive Materials Science Tetrahedron (MST) Approach in Long-Acting Injectable Suspensions Development

Zahra Salahshoor, Parind Desai, Sayantan Chattoraj, Aarti Gidh, Stephanie Lam

One of the approaches for manufacturing long acting injectable (LAI) suspensions is by wet bead milling, which reduces the particle size of API crystals in a liquid vehicle via grinding. A common failure mode observed during this "top-down" high energy process is long milling time to get to target particle size distribution. In order to avoid this failure mode, APIs with appropriate mechanical and physicochemical properties are needed to render the final suspension product in a feasible time. An early evaluation of the suitability of API is necessary to prevent delays in project timelines and accelerate the project progress. However, this may not be possible without an in-depth materials science tetrahedron (MST) approach, which connects the structure, properties, processing, and performance of an API. The MST approach leverages materials structure analysis and materials sparing tests, not only does it enable process development in a shorter time but also using less API and fewer experiments. Materials structure analysis and characterization reveal the properties of the material leading to choose the suitable form of the crystal earlier in the product development. Linking structures and properties to the process renders an appropriate manufacturing step and help establishing a robust control strategy at scale-up for consistent performance. In this study, compaction-based techniques and relevant analyses were used to measure the properties of APIs and correlate those to APIs' milling time as observed experimentally. Compaction simulator was used as a material sparing tool to develop a novel method of measuring Young's modulus at the bulk powder level. Bulk Young's modulus results explained the differences observed in the milling time span between different assets. Although the parameters directly impacting milling time were different some of the runs, a clear consistency was observed between the API's Young's Modulus and the time needed to achieve target size distribution, showing that Young's Modulus at bulk powder level is a good predictor for milling time. At all levels of drug product development, particularly during late-stage large-scale manufacturing, milling time span should be feasible to accommodate plants' operational capability, prevent bioburden risks, and other lengthy process time-related concerns. This mechanical characteristic can also be used to monitor lot to lot variability. In summary, the interdependency between crystal structure, properties, processing, and performance was studied. It is demonstrated that a systematic implementation of the MST approach would give us insight into the feasibility of wet bead milling unit operation at larger scales as well as expedite product development with limited amount of API and in compressed timelines.

654g Successful Shelf-Life Predictions of Amorphous Solid Dispersions

Gabriele Sadowski, Christian Lübbert

Amorphous solid dispersions (ASDs) are a state-of-the-art enabling formulation technique for poorly soluble small-molecule drugs. The bioavailability is significantly enhanced by the drug dissolution in a polymer carrier. The drugs' recrystallization in the polymer is highly unwanted as it marks the end of shelf life of such a formulation. The current work presents a novel fast-track prediction method that combines thermodynamic factors (water absorption, fundamental crystallization driving forces), kinetic factors (glass transition, diffusivity in the ASD), drug-specific crystallization properties (nucleation, crystal growth, glass-forming ability), thermal history (spray drying, melt extrusion, aging) and their mutual impact to predict the shelf life of an ASD at any drug load/ temperature/ humidity condition. This method is developed to predict ab-initio the crystallization onset time and thus the shelf life of metastable ASDs prior to storage tests and hereby dramatically reduces the risk of late-stage product failures due to stability issues. The shelf life predictions were validated by three-years-enduring stability tests of felodipine -containing ASDs. This approach serves as robust tool for predicting the shelf life of metastable ASDs in the entire temperature/drug load/-humidity space and allows a revolutionary risk assessment of storage test results at early stage of formulation development and even before conducting the tests. It was shown that stability data obtained at so-called 'accelerated', harsh storage conditions within short time do not at all reflect the shelf life at more moderate storage conditions, as thermodynamic and kinetic aspects differ completely. The developed model is highly flexible with respect to potential data input and provides a revolutionary approach for ASD shelf-life optimization in the future.

725a Leveraging COSMO-RS for Chemical Development and Scale-up in the Pharmaceutical Industry

Ines Rocha, Saúl Silva, Nuno Lousa

Chemical development and scale-up in the pharmaceutical industry are critical stages in the journey from drug discovery to commercialization. The efficient and robust transition from laboratory-scale synthesis to large-scale manufacturing is paramount for ensuring product quality, safety, sustainability, and cost-effectiveness. However, this process is often hindered by challenges such as limited experimental data, complex reaction pathways, and stringent regulatory requirements.[1] Indeed traditional approaches rely heavily on empirical experimentation, which is time-consuming, resource-intensive, and often limited by the availability of experimental data. Furthermore, the complexity of modern drug molecules and the stringent regulatory landscape necessitate innovative strategies for process optimization and formulation design.

In recent years, computational methods have emerged as indispensable tools for addressing these challenges and accelerating the development timeline.[2] Among these methods, COSMO-RS (Conductor-like Screening Model for Real Solvents) stands out for its ability to predict thermodynamic properties and phase behavior of complex chemical systems.[3] Rooted in quantum chemistry and statistical thermodynamics principles, COSMO-RS, offers a promising contribute for efficient chemical process development and scale-up. Unlike traditional empirical methods, it provides a molecular-level understanding of solvent-solute interactions, leveraging molecular descriptors and advanced algorithms to simulate complex chemical systems, offering insights into the underlying physicochemical phenomena governing solvation and phase behavior. [4-6]

The present work demonstrates COSMO-RS' capacity to develop robust distillations, liquid-liquid extractions, and crystallization processes through its accurate predictions of vapor-liquid equilibria (VLE), liquid-liquid equilibria (LLE) and solid-liquid equilibria (SLE), prediction of partition coefficients and distribution ratios for target solutes and estimation of solubility data.

Overall, a successful assessment of Cosmo-RS for simulating these unit operations was accomplished, combining theoretical data, experimental validation, and practical applications with very positive impact on operational safety, cycle time and resource management.

References

[1] Federsel, H. Chemical Process Research and Development in the 21st Century: Challenges, Strategies, and Solutions from a Pharmaceutical Industry Perspective. *Acc. Chem. Res.* **2009**, 42, 5, 671–680

[2] Rogers, A.; Ierapetritou, M. Challenges and opportunities in modeling pharmaceutical manufacturing processes. *Comput. Chem. Eng.* **2015**, 81, 32-39

[3] Klamt, A. The COSMO and COSMO-RS solvation models. *Wiley Interdiscip. Rev. Comput. Mol. Sci.* **2017**

[4] Klamt, A. Prediction of the mutual solubilities of hydrocarbons and water with COSMO-RS. *Fluid Phase Equilibria*, 206 (1), **2003**, 223-235

[5] Klamt, A. COSMO-RS for aqueous solvation and interfaces, *Fluid Phase Equilibria*, **2016**, 407,

152-158

[6] Freire, M.; Santos, L.M.N.B.F.; Marrucho, I.M.; Coutinho, J.A.P. Evaluation of COSMO-RS for the prediction of LLE and VLE of alcohols + ionic liquids. *Fluid Phase Equilibria*, **2007**, 255 (2), 167-178

[7] Putnam, R.; Taylor, R.; Klamt, A.; Eckert, F.; Schiller, M. Prediction of Infinite Dilution Activity Coefficients Using COSMO-RS. *Ind. Eng. Chem. Res*. **2003**, 42, 3635-3641

725b Prediction of Thermal Decomposition Hazards for Pure Compounds: A Machine Learning Approach Utilizing Molecular Structure

Subhadra Devi Saripalli, Rajagopalan Srinivasan

Identifying potential hazards during the initial stages significantly enhances the safety of the chemical manufacturing process. Thermal decomposition is a highly critical reactive hazard and has been a major cause of several industrial disasters such as the T2 laboratory explosion[1] and BP Aramco polymers explosion[2]. Thermal onset temperature is the crucial property of a chemical compound that determines the probability of thermal decomposition. Onset temperature is typically obtained through calorimetric experiments or Density Functional Theory (DFT) simulations. The experimental approach is often hindered by the limited availability of reaction mass for comprehensive thermal hazard testing during the initial stages, while DFT calculations are computationally expensive. Therefore, an alternative approach to predict the onset temperature without experimentation and expensive computational DFT simulations is needed. However, onset temperature is a property of a chemical dependent on its molecular structure and defines the stability of the compound to heat. In this work, we aim to develop a machine learning (ML) model capable of predicting thermal onset temperature by correlating it with molecule structural features.

Despite challenges such as data limitations and complex reaction mechanisms, ML algorithms offer a transformative solution for predicting thermal hazards and safety-related properties with reasonable accuracy. Several models have been developed to predict thermal decomposition parameters for various classes of compounds using their structural information. For instance, Fayet et al. (2011)[3] developed a QSPR model by integrating Multiple Linear Regression (MLR), Partial Least Squares (PLS), and decision tree models to predict heat of decomposition for nitroaromatic compounds. Zhang and Xu (2020)[4] employed a Gaussian process regression model to predict the decomposition onset temperature of lubricant additives. He et al. (2021)[5] utilized a genetic algorithm and multiple linear regression model to predict the thermal decomposition temperature of binary Imidazolium ionic liquid mixtures. This literature provides insights into the feasibility of predicting thermal decomposition parameters for specific compound classes. These models help to understand the stability of similar compounds, aiding in identifying potential thermal hazards involved in the processes.

The methodology proposed for developing the model for predicting the onset temperature using the molecular structure involves collecting the data, clustering appropriately, and developing a model using various descriptors. The onset temperature prediction model developed in this work uses the data available in the literature. By limiting the dataset to pure compounds, a maximum of up to two rings, and the presence of reactive functional groups such as amines, alcohols, nitro, and azo groups, approximately 1000 data points were selected. Molecular representations in the form of a Simplified Molecular Input Line Entry system (SMILES) and the physicochemical and topological descriptors were extracted. A generalized model may not yield accurate results and cause underfitting because structural changes significantly affect the onset temperature. Therefore, clustering the data based on thermal behavior and structural similarity was performed. The cluster with sufficient data, namely nitro compounds, was chosen for model development. The most suitable descriptors were identified using various feature selection methods and incorporated into the modeling. A multiple linear regression model was developed using the selected descriptors. The developed model undergoes testing and validation, involving multiple iterations to fine-tune model configurations, ensuring its robustness and enhancing its predictive capabilities. The developed model is aimed to attain optimal accuracy with the mean absolute error (MAE) of less than 20.

This model offers insights into thermal onset temperature at an early stage without experimentation or computationally expensive DFT simulations. Prediction of onset temperature suggests the criticality of the process by comparing the onset temperature with operating conditions. This enhances safety analysis by identifying hidden thermal hazards early in the process, reducing resource wastage, and optimizing scale-up potential.

Keywords: Thermal decomposition, Onset temperature, molecular descriptors, predictive modeling

References

(1) *T2 Laboratories Inc. Reactive Chemical Explosion | CSB.* https://www.csb.go12v/t2-laboratories-inc-reactive-chemical-explosion/ (accessed 2024-04-04).

(2) *BP Amoco Thermal Decomposition Incident | CSB.* https://www.csb.gov/bp-amoco-thermal-decomposition-incident/ (accessed 2024-04-04).

(3) Fayet, G.; Del Rio, A.; Rotureau, P.; Joubert, L.; Adamo, C. Predicting the Thermal Stability of Nitroaromatic Compounds Using Chemoinformatic Tools. *Molecular Informatics* **2011**, *30* (6–7), 623–634. https://doi.org/10.1002/minf.201000077.

(4) Zhang, Y.; Xu, X. Machine Learning Decomposition Onset Temperature of Lubricant Additives. *J. of Materi Eng and Perform* **2020**, *29* (10), 6605–6616. https://doi.org/10.1007/s11665-020-05146-5.

(5) He, H.; Pan, Y.; Meng, J.; Li, Y.; Zhong, J.; Duan, W.; Jiang, J. Predicting Thermal Decomposition Temperature of Binary Imidazolium Ionic Liquid Mixtures from Molecular Structures. *ACS Omega* **2021**, *6* (20), 13116–13123. https://doi.org/10.1021/acsomega.1c00846.

725c Transport-Kinetic Modeling of a Double N-Debenzylation in the Production of an Active Pharmaceutical Ingredient

Ted Raymond, Thomas Toupy, Yihui Yang, Raymond Turro, Neda Nazemifard, Charles D. Papageorgiou

Protection of heteroatoms by benzyl groups is a common and effective strategy in organic synthesis. Subsequent deprotection is commonly performed by catalytic hydrogenolysis. Published literature focuses primarily on O-debenzylations; however, N-debenzylations are also reported.[1] Whereas O-debenzylations are commonly reported to be fast, many reports of N-debenzylations show slower kinetics. This is attributed to the relative differences in bond polarity.[2] The system under investigation involves hydrogenolysis of an N,N-dibenzyl amine on a Pd/C catalyst.

During the synthesis of an intermediate compound, it was observed that the reaction completion time in lab-scale experiments and manufacturing was highly variable at nominally identical conditions. At nominally identical conditions, the reaction was sometimes observed to complete in a matter of hours, and sometimes as long as several days. To diagnose the issue, a thorough exploration and characterization of the reaction space was performed, and a process model was developed.

Since hydrogenation is a multiphase reaction, the initial process development efforts were focused on assessing the gas-liquid mass transfer across the scales. To better understand these effects, pressure-drop experiments were performed in multiple reactors across lab and manufacturing scales. The overall mass transfer coefficients were regressed based on the dynamics of the reactor headspace pressure decrease. To improve the correlation and mapping, CFD simulations were also performed to generate estimates of the mass transfer coefficients.

As direct measurement of hydrogen composition is nontrivial, we next aimed to improve our estimation of the hydrogen partial pressure and thermodynamic solubility. The NIST-modified UNIFAC activity coefficient model[3] was employed to improve the calculation of the vapor pressure contributions from liquid-phase species. As there was also some uncertainty of the effects the reaction species may have on hydrogen solubility, we performed a comprehensive survey of the literature for published data. The compiled data spanned the key functional groups present in our system. Using this, molecular hydrogen was treated as a new group, and corresponding UNIFAC parameters were regressed.

An additional challenge presented itself after a substantial amount of kinetic data had been collected. In experiments where the reaction completed quickly, subsequent overreaction of the debenzylated product was frequently observed. Interestingly, these impurities were not observed to grow in experiments that were slower to complete, even when held for long after the completion of the reaction. A detailed reaction kinetic model is developed by taking into account the complex competitive adsorption and variation in molecular sizes based on extended LeVan-Vermeulen adsorption isotherms[4]. A combination of kinetic model, CFD simulations, and mass transfer analysis were used to assess the sensitivity of reaction time across lab and manufacturing scales and to provide a robust scale up strategy.

References

1. David, A., & Vannice, M. (2006). Control of catalytic debenzylation and dehalogenation

reactions during liquid-phase reduction by H2. *Journal of Catalysis*, *237*(2), 349–358. https://doi.org/10.1016/j.jcat.2005.11.017

2. Smith, G. V., & Notheisz, F. (1999). Chapter 4 - Hydrogenolysis. In *Heterogeneous Catalysis in Organic Chemistry* (pp. 119–218). Academic Press.
3. Kang, J. W., Diky, V., & Frenkel, M. (2015). New modified UNIFAC parameters using critically evaluated phase equilibrium data. *Fluid Phase Equilibria*, *388*, 128–141. https://doi.org/10.1016/j.fluid.2014.12.042
4. Frey, D. D., & Rodrigues, A. E. (1994). Explicit calculation of multicomponent equilibria for ideal adsorbed solutions. *AIChE Journal*, *40*(1), 182–186. https://doi.org/10.1002/aic.690400121

725d Scale up Methodology and Practical Application of Wet Bead Milling for the Production of Nano and Micro Sized Particles

Christopher J. Morrison

Particle size reduction via wet bead milling of pharmaceutical materials to a reproducible nano and micro sized particle endpoint is achievable across the Netzsch Delta Vita mill series. The controlled particle size allows for a sustained long-acting release of the active pharmaceutical ingredient to a patient to reduce the number of doses in a given year. Wet bead milling offers the advantage of increased heat dissipation and removal for temperature sensitive products when compared to a dry milling process. A total energy model approach was used to properly scale the process while geometrically scaling equipment and holding key process variables including formulation composition, agitator tip speed, and bead characteristics, among other parameters, constant between scales. An advantage of scaling with constant tip speed is that the power and heat transfer surface area also scale by the diameter of the mill which results in similar cooling across scales thereby avoiding unwanted surprises for temperature sensitive products during scale up. The milling process was successfully scaled from lab to industrial scale. A case study will be presented showing the experimental design, scale-up calculations, and results.

725e Credibility Assessment of Mechanistic Roller Compaction Model Using ASME Vandv 40 Risk Informed Framework

Santhosh Seshadhri, Marc Horner, Senthil Kumarasamy, Pankaj Doshi

Roller compaction (RC) is a continuous dry granulation process in which the primary powder is compressed at high pressure by two counter rotating rollers to produce a ribbon. The ribbon is then crushed to produce granules. This prevents segregation of the constituents of the powder mixture and improves the flow properties and compaction characteristics. However, RC implementation for high drug load formulations can be challenging due to flow issues and a high consumption of active pharmaceutical ingredient (API). This has led to significant growth in the use of predictive mechanistic models in the design and scale-up of roller compaction processes over the past decade.

RC mechanistic model utilizes the analytical method developed by Johanson [Johanson, J. R. (1965)] for predicting compaction process performance. Their model predicts ribbon solid fraction as a function of roll gap, roll width and roll force parameters. The ribbon solid fraction is an important intermediate parameter that the process development team relies upon as an indicator of overall tablet quality. This provides guidance on the starting point of process operating conditions which will then be utilized during tech transfer activities, thus saving cost and time during drug product development.

It is important to establish that the model is credible for its intended use given the criticality of the RC model predictions for process development. Currently, there is no consensus on how to establish and assess model credibility, including the selection of appropriate verification and validation activities for pharmaceutical process models.

In this work, we summarize the application of the risk informed credibility framework outlined in the ASME Verification & Validation 40 (ASME V&V 40) standard for the roller compaction mechanistic model credibility evaluation (American Society of Mechanical Engineers, 2018). This standard provides a risk-informed credibility assessment framework for the development and assessment of computational models of medical devices, which was later adopted into model-informed drug development (Kuemmel et al., 2020; Viceconti et al., 2021). The key aspects of the framework include defining the Question of Interest that the model is going to address, the Context Of Use of the model, assessing model risk based on the impact of the model's influence in the decision as well as model's impact on product quality and how this affects the patient, and establishing model credibility through various verification and validation methods described in ASME V&V 40 framework. This presentation will review the application of the ASME V&V 40 standard to establish the credibility of the RC to support scale-up activities.

725g Predictive Drying Recipe Development for Drying a Heat-Sensitive and Sticky Powder: Scale-up from Lab to Manufacturing By Modeling and Simulation

Nima Yazdanpanah

Residual solvent content of powdered product from crystallization/filtration or spry drying is a quality aspect that related to CQA and ICH threshold. The residual solvent down to ppm level may cause form change, stability, and impose toxicity or bioavailability alteration. The secondary dryers in types of double cone dryers, agitated mixing dryers, fluidized bed, rotary and so are common unit operations in the industry. The transport phenomena involved in the drying of a power bed is very complex and incorporate the heat and mass transfer and mixing, which all are scale dependent. The scale dependency becomes a challenging problem in scale-up and technology transfer, where hundreds of kilograms of expensive and sensitive material should be processed. Under- or over-processing of the material may cause thermal degradation (impurity formation), agglomeration or breakage (change of PSD), or solvent-mediated phase change. The conventional scale-up method is still experimental trial and frequent sampling at the large manufacturing scale. This is a very risky and expensive [blind] process that is developed based on no knowledge of the material or equipment or process. In contrary, mechanistic modeling can be used to simulate all the involved transport phenomena in a predictive fashion.

This presentation will demonstrate a manufacturing scale drying process development and prediction for an Ekato 1600 SolidMix dryer with the capacity of 800kg, for drying a sticky and heat sensitive material to reduce Methanol and Dichloromethane from percentage level to PPM over 50 hours. The diverse heat transfer models, diffusion and evaporation, inner-bed mixing, stickiness, and residual solvent content trend will be discussed for a range of process parameters scenarios. The models are multiscale, multidomain, and Multiphysics with different levels of fidelity and computation expenses.

725h Leveraging Mother Liquor Viscosity in Non-Ideal Systems to Optimize Filtration Time

Rachel Janes, Vladislav Lisnyak, Christopher Jordan, Kenneth Fraunhoffer

In pharmaceutical processing, filtration is a key unit operation in the production of active pharmaceutical ingredients (API) and API intermediates. Long filtration times can greatly impact cycle time and filtration issues can be more complex when using a centrifuge. To prepare for scale up into a centrifuge, understanding the parameters which can influence the filtration performance (i.e. flux, compressibility) is key to predicting cycle times. Many factors contribute to filtration performance, but one factor that is often overlooked is viscosity of the mother liquor. N-Methylpyrrolidone (NMP) is a common polar process solvent, however NMP/ water and NMP/ alcohol systems are non-ideal and result in very viscous mixtures that can negatively impact filtration time. A case study will be presented where the filtration of a pharmaceutical intermediate from an NMP/water system was improved through consideration of mixture viscosity. The development included an assessment of solvent selection and crystal property modifications. Filtration flux measurements and process modeling were used to identify the benefit of a ternary solvent system to reduce supernatant viscosity. This resulted in an improved filtration flux while maintaining product yield and quality.

89a Advancing the Narrative Around Predictive Stability – a Kinetic Study of Ophthalmic Liquid Formulation Stability for Pilocarpine Solutions

Eric Sacia, Fang Wang, Sandeep Kunati, Russell Hertzler

While both data-driven and first principles models can be useful in describing chemical phenomena, the rooting of first principles models in the fundamental physics that describes chemical reactions can allow for both more predictive and more accurate extrapolation of a formulation's inherent chemical stability. In this talk, we will discuss the stability of pilocarpine in aqueous solutions. Pilocarpine is a muscarinic acetylcholine agonist and is the active pharmaceutical ingredient (API) in ophthalmic liquid formulations used in conditions such as presbyopia, increased intraocular pressure, and angle-closure glaucoma. To successfully formulate this API into a doseable liquid ophthalmic formulation, building an understanding into its chemical stability is essential. Further, due to the extended cycle time of performing stability studies at representative storage conditions, it is highly desirable to build an understanding of how the formulation's stability may progress using an accelerated stability program.

Performing targeted stability studies at elevated temperatures in representative container closures provides an opportunity to understand the inherent chemical pathways that lead to pilocarpine degradation, particularly through epimerization to form isopilocarpine and lactone hydrolysis to form pilocarpic acid. These chemical transformations also impact other CQA's, such as the solution's pH, over time. Further, by comparing closures, we assessed the impact of mass transfer through the packaging via diffusion as well as looking at potential surface effects associated with the chosen packaging for the stability study. By building a thorough understanding of the mechanisms that promote pilocarpine degradation, a highly predictive model was generated that accounts for the autocatalytic decomposition of pilocarpine as well as the solution's pH over time in the aqueous formulation. This model, which was trained on several months of accelerated stability conditions, could then be used to predict up to three years of ambient temperature stability with very good performance.

89b Reaction Kinetics Modeling of a Challenging Amide Coupling Step: Quantitative Process Understanding and Enhanced Robustness

Jingfan Yang, Adriana Jemison, Audrey Mack

This talk will describe the development and application of a reaction kinetics model of an amide coupling step, which, due to the convergent nature of the synthesis and limited impurity rejection downstream, proved challenging. Incorporating the dissolution/desupersaturation kinetics of the poorly soluble starting materials, as well as a competitive hydrolysis side reaction, the model (1) identified the reactants' stoichiometric ratio as a critical parameter and pinpointed the most reliable process center point, (2) evaluated *in silico* the process robustness improvement from proposed in-process control changes, (3) corrected prior assumptions of the phase equilibria, and lastly, (4) confirmed the causality between atypical experimental observations and poor moisture control, helping to establish KF specifications of the reagents. Throughout the process development, the modeling and experimental efforts guided, refined, and validated each other synergistically. In summary, the reaction kinetics modeling enabled quantitative process understanding and enhanced the process robustness of a challenging step.

89c A Comprehensive Modeling for Predicting Oxygen-Related Degradation in Pharmaceutical Drug Product Stability

Kyu Yoon, Zhibin Tan, Prachi Gupta, Laura Mozdzen, Vivek Narsimhan, Sherwin Shang

A comprehensive model integrating mass transfer and reaction kinetics of a drug product in a primary container to predict pharmaceutical drug product stability was developed. The study was focused on oxygen-induced degradation during product shelf life. Traditional accelerated stability study and analysis, although efficient in gathering data quickly, may not accurately represent the behavior of drug products during the real-life storage conditions.

Our research highlights the potential inadequacies of assumptions made by well-mixed models, particularly when the real-life environment transitions to regimes are limited by reaction kinetics. By incorporating the Damköhler number, our integrated model can effectively bridge the gap between drug product conditions used in accelerated tests and those encountered in real-life storage. Damköhler numbers are dimensionless numbers used to connect the chemical reaction timescale (reaction rate) to the mass transport rate occurring in a system.

The findings indicate that the conventional method of extrapolating kinetic constants from accelerated tests can lead to significant overestimations, especially in reaction-limited scenarios. The model, validated with experimental data, offers improved accuracy in predicting the long-term stability of drug products, surpassing the capabilities of well-mixed models. This study underscores the importance of developing reliable predictive models that account for the significant roles of diffusion and permeation under real-life storage conditions, ensuring the sustained efficacy and accuracy of pharmaceutical drug product stability throughout the shelf life.

89e Process Knowledge for Drug Substance Production Via Kinetic Modeling, Parameter Estimability Analysis and Reaction Optimization

Iman Moshiritabrizi, Jonathan McMullen, Brian M. Wyvratt, Kimberley B. McAuley

A fundamental model is developed to study the formation of 2,6-difluoropurine-9-THP from starting material 2,6-dichloropurine-9-THP. The 2,6-difluoropurine-9-THP product is an intermediate used in the synthesis of Islatravir (MK-8591), a therapy for treatment of HIV. The corresponding reaction scheme is shown in Scheme 1 where the starting material is labeled (SM).[1,2] Trimethylamine (TMA), a gaseous material, is bubbled into the liquid solution in the reactor to start the first reaction. Kinetic parameters for the first reaction and subsequent reactions are estimated from 26 batch reactor experiments. An Error-in-Variables-Model (EVM) approach is used for parameter estimation to address uncertainty in initial concentrations of TMA. Uncertainty in the TMA charge stems from the combined variability in the feed flowrate of TMA and in the TMA charge time. The EVM approach in the current study ranks the combined set of parameters and uncertain inputs from most estimable to least estimable. Then it estimates the true values of the uncertain inputs along with the model parameters. Finally, a mean-squared-error criterion is used to determine the appropriate parameters and inputs that should be estimated, based on the ranked list to avoid overfitting.[2] This parameter subset selection method is used to determine that 33 out of 39 model parameters should be estimated along with 26 uncertain initial concentrations. The remaining six parameters are kept at their initial values to prevent overfitting of available data. EVM parameter estimates are compared with estimates obtained using a traditional weighted-least-squares (WLS) approach that neglects input uncertainties. The EVM estimates provide a better fit to the data and, as shown using cross-validation, improved accuracy for model predictions. The resulting model and EVM parameter values are used to find reactor conditions that maximize product yield while obeying constraints on temperature, the initial ratio of TMA to starting material, batch time, and the volume of solvent. An optimal yield of 92.04 % is predicted, which is higher than the yield of 90.26 % at the best experimental conditions in the data set. Contour plots are used to highlight the insensitivity of the optimal yield to batch time and solvent volume, indicating that a yield of 91.83 % could be obtained using a 50 % lower batch time and 33 % less solvent.

Scheme 1. Reaction scheme used to produce 9-THP-2,6-difluoropurine, which is labeled (P) in the black box

References

1. Hong CM, Xu Y, Chung JY, et al. Development of a commercial manufacturing route to 2-fluoroadenine, the key unnatural nucleobase of islatravir. Organic Process Research & Development. 2020;25(3):395-404.
2. Moshiritabrizi I, Abdi K, McMullen JP, Wyvratt BM, McAuley KB. Parameter estimation and estimability analysis in pharmaceutical models with uncertain inputs. AIChE Journal. 2023:e18168.

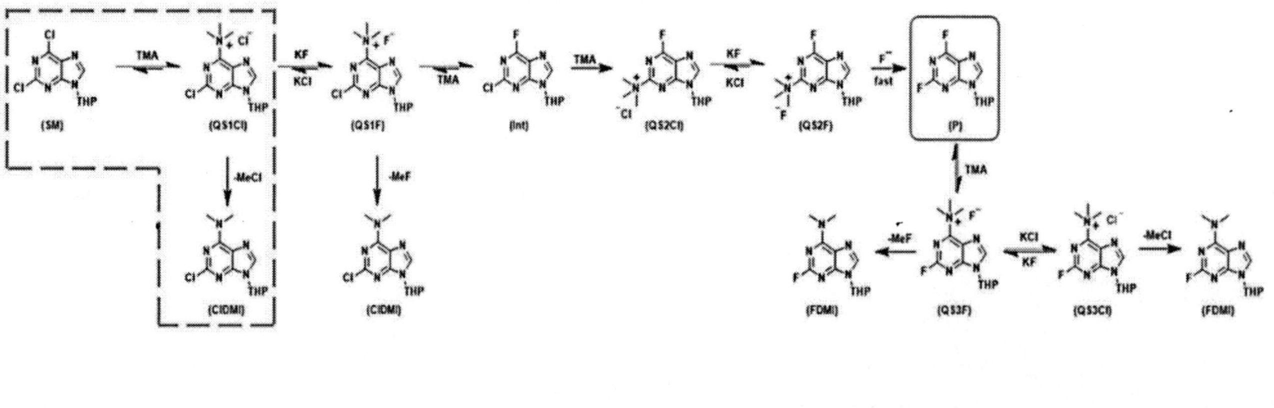

89f High Pressure Reaction Kinetic Modeling: Leveraging Single Data-Rich Experiment

Ines Rocha, Pedro Jorge, Nuno Lousa, Saúl Silva

High pressure reactions play a pivotal role in various industries, ranging from pharmaceuticals to petrochemicals, where the manipulation of pressure conditions can significantly alter reaction kinetics, selectivity, and product yields.[1]

Safety should not be taken lightly when considering high pressure reactions. A number of accidents has already shown how dangerous these types of reactions can be [2]. Another challenge in high pressure reactions is the potential for side reactions or unexpected reaction pathways under extreme pressure conditions. These reactions can result in the formation of undesirable secondary products, reduced selectivity and decreased overall process efficiency. Understanding the underlying chemistry and reaction mechanisms is important to mitigate the risk associated with these challenges and optimizing reaction conditions to maximize productivity. Performing high pressure reactions additionally creates operational difficulties, namely during sampling, illustrating the urge to implement more PAT tools and models to reduce operational burden.

In this study, an efficient approach for developing kinetic models for high-pressure reaction systems is described. Substantial value of employing just one data-rich experimental reaction allied with computational capability is demonstrated. Leveraging computational expertise, a digital twin was developed, comprising kinetics of a high-pressure reaction and Vapor-Liquid Equilibrium (VLE), with a key emphasis on temperature-dependent behavior.

(Figure 1. Process workflow from a digital twin to a complete model)

Described methodology aimed to build a kinetic model with just one reaction experiment. Through advanced computational systems, this single reaction becomes the cornerstone for adjusting vital kinetic parameters, such as rate constant and activation energy. Furthermore, to fully incorporate temperature dependency in the model, detailed influence on generated pressure was achieved with good accuracy. This temperature-dependent digital twin stands out by its predictive capability of simulating reaction behavior across a wide range of operating temperatures.

Due to reaction and experimental setup complexity, discrete sampling during reaction was challenging. Alternatively, in-line infrared (IR) spectroscopy was employed as Process Analytical Technology (PAT) tool, using ReactIR from Mettler Toledo. Prior to the reaction, an IR calibration curve was performed to determine concentration of starting material, 2-(methylthio)pyrimidin-4(3*H*)-one, and measure consumption during reaction execution.

(Figure 2. Reaction Scheme)

Reaction profile was previously predicted using available kinetic parameters of the same reaction with a different substrate. Reaction was slower than predicted, but after sequential temperature increases, higher reaction rates were observed. The kinetic constant was thereafter fitted, exhibiting a twenty-fold decrease compared to the initial proposed value. This effect was somewhat anticipated based on the structural differences between the studied molecule and the one used as reference. Following this experiment, the next step is to forecast the reaction at a desirable isothermal temperature and validate the model.

(Figure 3. Temperature and ReactIR profiles during reaction)

Our methodology represents an unconventional approach, demonstrating that a single experimental reaction coupled with computational capabilities can yield comprehensive insights into the kinetics of reactions, namely under pressure. This streamlined approach not only minimizes experimental resources but also accelerates the development of kinetic models, offering unprecedented efficiency and accuracy in understanding complex reaction systems.

Figure 1. Process workflow from a digital twin to a complete model

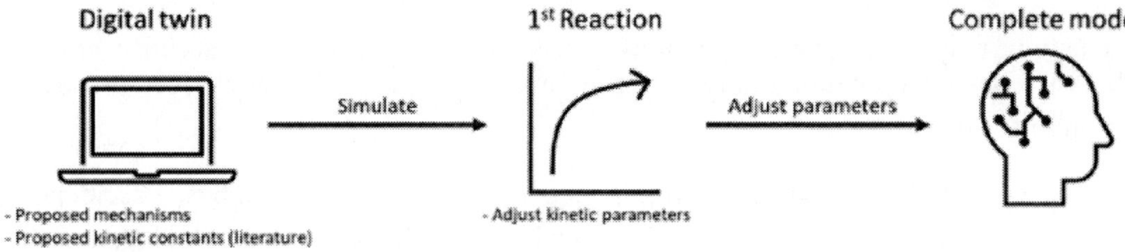

Figure 2. Reaction Scheme

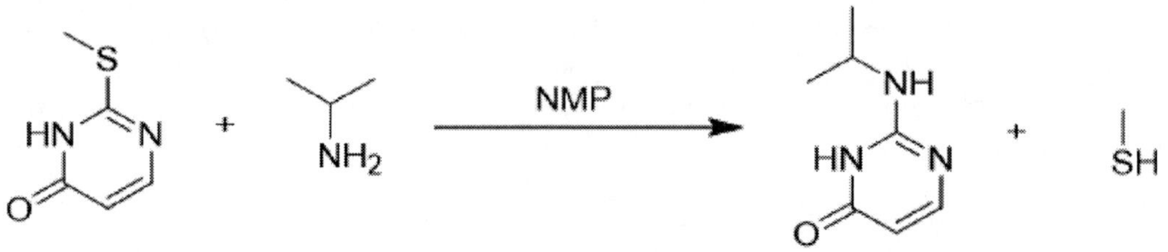

Figure 3. Temperature and ReactIR profiles during reaction

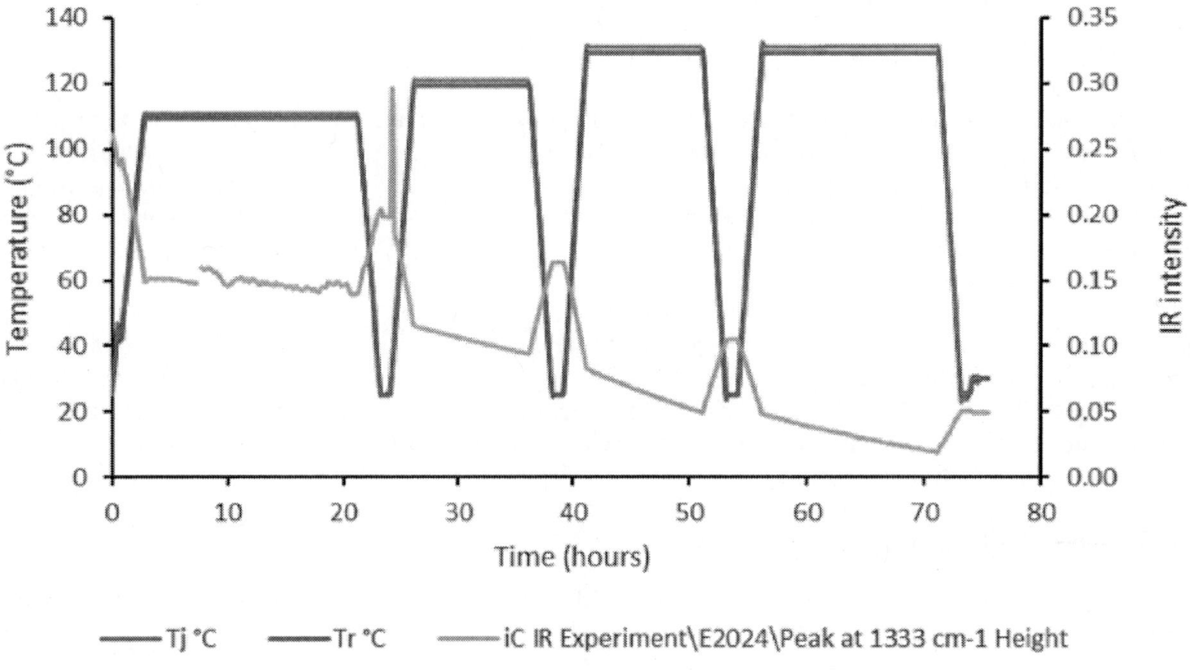

89g Modeling API Crystal Surface Degradation Processes in Drug Product Formulations

Patrick Corona, Benjamin M. Cohen, Jose Tabora

Modeling the packaged stability of active pharmaceutical ingredient (API) drug product (DP) is critical for predicting the shelf life of DP formulations. The most common modeling approach is to assume the DP degradation rate is only a function of the temperature and relative humidity (RH). The temperature and RH dependencies determined from experiments are then combined with a model for package moisture uptake to predict long term shelf stability. Recently, an API in development was discovered to have a degradation rate that depends on the specific surface area (SSA) of the API drug substance (DS), suggesting that the degradation mechanism is driven by processes on the surface of the API. To capture this experimental observation, a mechanistic model for DP degradation due to surface reactions was developed. Open dish stability experiments were leveraged to fit the model and comparisons between model predictions and packaged stability experiments validated the model. Model predictions were then leveraged to inform decisions including the choice of desiccant loading in bottles and API DS impurity and SSA specifications. Finally, potential improvements to the model, including application to other SSA-dependent degradation processes, will be discussed.

88a Biologics Application CR Trend Analysis and Advancing Regulatory Convergence and Reliance Through a Pharmaceutical Quality Knowledge Management Network (invited talk)

Stelios C. Tsinontides

The presentation will provide a brief analysis of the causes of Complete Responses (CRs) to recent original and Biosimilar BLA applications, including the impact of applications using CMOs, and factors cited in facility inspections. The presentation will also include an update on ICMRA's Pharmaceutical Quality Knowledge Management project, including an overview of the learnings from two pilot programs on collaborative assessment of post-approval changes and hybrid inspections by multiple authorities. FDA has been leading the effort with other Regulators and Industry to use two pilot programs to understand opportunities and challenges in developing a Global submission pathway, enabling Reliance in CMC assessment and facility inspection, and thus reduce and even eliminate need for duplicative Regulatory submissions and inspections. Given the increased use of advanced and innovative technologies to speed up the development of medicines and the limited regulatory resources, our efforts are aimed to ensure the most appropriate deployment and use of tools and knowledge to speed up regulatory reviews to make safe and high-quality drugs available to patients on a global basis.

88b An Examination of Process Models and Model Risk Frameworks for Pharmaceutical Manufacturing (invited talk)

Thomas O'Connor

Process models are a growing tool for pharmaceutical manufacturing process design and control. The Industry 4.0 paradigm promises to increase the amount of data available to understand manufacturing processes and tools such as AI and machine learning that might accelerate process development and allow better prediction of process trajectories. Several examples of the application of process models have shown process improvements can be realized in lyophilization, chromatography, fluid bed drying, bioreactors, and continuous direct compression and wet granulation. An important consideration of implementing a process model is determining the impact of the model on quality of the product and the risks associated with model maintenance over product lifecycle. Several documents address these risk-based considerations, including: (a) the ICH ICH Quality Implementation Working Group Points to Consider (R2), (b) the FDA guidance for industry on Development and Submission of Near Infrared Analytical Procedures, (c) the EMA guideline on the use of near infrared spectroscopy by the pharmaceutical industry and the data requirements for new submissions and variations, (d) ICH Q13: Continuous Manufacturing of Drug Substances and Drug Products and (e) the American Society Mechanical Engineers Verification and Validation 40 standard. In this presentation, hypothetical case studies are to illustrate the implications of applying a framework to determine model risk to facilitate model validation and lifecycle maintenance in the manufacture of pharmaceuticals and biological products.

88c Elucidating the Binding Mechanism of Paramagnetic Amyloid Ligands to α-Synuclein Monomer

Hung-Yu Wan, Surl-Hee Ahn

It is well known that the hallmark of Parkinson's disease (PD) is the accumulation of α-synuclein oligomers (αSOs), which is one of the aggregate forms of α-synuclein. Recently, our collaborator, Prof. John Voss (UC Davis), has found that paramagnetic amyloid ligands (PALs) could potentially inhibit the formation of αSOs. However, the specific mechanism of how PALs interact and block the aggregation is unknown, which is crucial for drug discovery for PD.

In this study, we utilized Ligand Gaussian accelerated Molecular Dynamics (LiGaMD) simulations to examine the interactions between PALs and the α-synuclein monomer. Our findings indicate that a 1μs LiGaMD is sufficient for our system, as evidenced by the near-zero auto-correlation values observed after the 1 μs run. Notably, we discovered that most PALs exhibit the highest molecular dynamics (MD) contact probability within the non-amyloid β component (NAC) region, which spans residues 61 to 94 of the α-synuclein monomer. This suggests that PALs are more likely to bind to the NAC region instead of the N-terminal or C-terminal regions of the α-synuclein monomer. Additionally, our results suggest that PALs may impede the formation of secondary structures, particularly α-helices and β-sheets. Moreover, to identify the most predominant structure within our system, we utilized k-means clustering analysis, which shows that there is not much difference between the bound and unbound states of the α-synuclein monomer.

88d A Knowledge-Graph-Based Pharmaceutical Engineering Chatbot for Drug Discovery

Naz Pinar Taskiran, Chia-En Tsai, Arijit Chakraborty, Vipul Mann, Venkat Venkatasubramanian

Despite their great success in day-to-day applications, ChatGPT and other large language models (LLMs) have not covered as much ground in scientific and engineering domains. One key challenge is the abundance of domain-specific terminology, which an LLM is not trained to extract in accordance with the underlying physical laws. This can lead to unreliable results or hallucinations. To address these challenges, we have developed SUSIE, an ontology-based pharmaceutical information extraction tool that is built to extract semantic triples and present them to the user as knowledge graphs (KGs) [1,2]. While KGs help visualize the relationships between different entities, they are not easily accessible for user questions, yet serve as structured inputs for LLMs. Thus, KGs can be used to efficiently query a corpus of pharmaceutical documents, streamlining drug discovery and manufacturing processes. In this work, we present a customized question-and-answering module that enables the user to query from generated KGs, and get an answer in natural language. We show that with the integration of prompt engineering into a model based on the set of rules defined by the ontology, the output is sensible and explainable.

Bibliography:

(1) Mann V., Viswanath S., Vaidyaraman S., Balakrishnan J., Venkatasubramanian V., (2023).

SUSIE: Pharmaceutical CMC ontology-based information extraction for drug development using machine learning,

Computers & Chemical Engineering, Volume 179, 108446.

https://doi.org/10.1016/j.compchemeng.2023.108446.

(2) Remolona M. F. M., Conway M. F. , Balasubramanian S., Fan L., Feng Z., Gu T., Kim H., Nirantar P. M. , Panda S., Ranabothu N. R. , Rastogi N., Venkatasubramanian V., (2017).

Hybrid ontology-learning materials engineering system for pharmaceutical products: Multi-label entity recognition and concept detection.

Computers & Chemical Engineering, Volume 107, Pages 49-60.

https://doi.org/10.1016/j.compchemeng.2017.03.012.

88e Exploring the Interlinks between SOS Response, Mutagenesis, and Resistance during the Recovery Period

Sreyashi Ghosh, Mehmet Orman

Although the mechanistic connections between SOS-induced mutagenesis and antibiotic resistance are well-established, our current understanding of the impact of SOS response levels, recovery durations, and transcription/translation activities on mutagenesis remains relatively limited. In this study, when bacterial cells were exposed to mutagens like ultraviolet light for defined time intervals, a compelling connection between the rate of mutagenesis and the RecA-mediated SOS response levels became evident. Our observations also indicate that mutagenesis primarily occurs during the subsequent recovery phase following the removal of the mutagenic agent. When transcription/translation was inhibited or energy molecules were depleted at the onset of treatment or during the early recovery phase, there was a noticeable decrease in SOS response activation and mutagenesis. However, targeting these processes later in the recovery phase does not have the same effect in reducing mutagenesis, suggesting that the timing of inhibiting transcription/translation or depleting energy molecules is crucial for their efficacy in reducing mutagenesis. Active transcription, translation, and energy availability within the framework of SOS response and DNA repair mechanisms appear to be conserved attributes, supported by their consistent manifestation across diverse conditions, including the use of distinct mutagens such as fluoroquinolone antibiotics and various bacterial strains.

Furthermore, we implemented a two-step screening methodology to gain a deeper understanding of the downstream mechanisms of RecA-mediated mutagenesis. In the first method, we screened RecA-LexA-regulated promoters by exposing cells carrying the promoter reporters to UV, aiming to identify SOS response genes significantly upregulated upon UV exposure (Fig.1). In the second method, we screened an *E. coli* knockout cell library encompassing all potential repair mechanisms that might be involved in the observed mutagenesis, particularly focusing on genes whose expression was not significantly upregulated in our initial screening. Through these methods, we identified several genes whose deletions significantly affected either mutagenesis alone (e.g., *umuC* and *umuD*) or both mutagenesis and cell culturability (e.g., *uvrA* and *recB*) (Fig.1). Monitoring the expression level of the candidate genes using flow cytometry provided a deeper insight into their regulation and potential roles in UV-induced mutagenesis and cell culturability. Overall, this study sheds light on the complex interplay among the convoluted, error-prone mutagenic pathways in the context of antibiotic resistance and evolutionary processes.

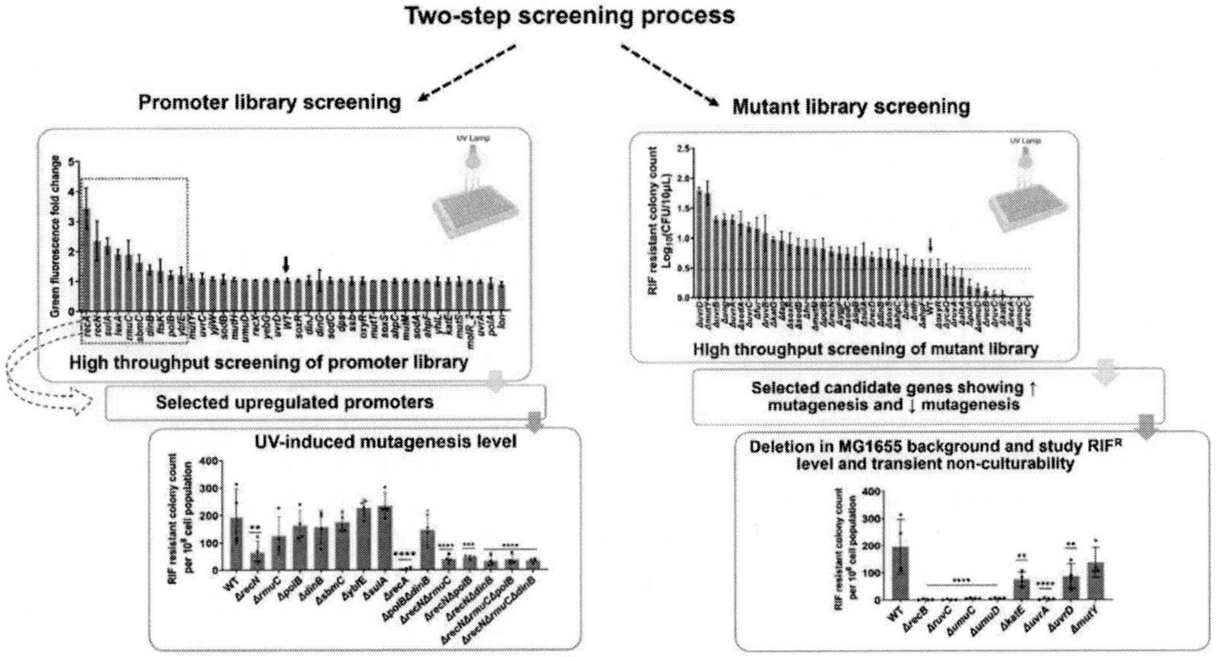

Fig. 1: UV-induced upregulation of SOS genes and their effects on mutagenesis. A two-step screening process was employed to investigate the underlying mechanism of RecA-mediated mutagenesis. First, a library of *E. coli* MG1655 strains with promoter reporters was treated by UV for 16 min in the mid-exponential phase and recovered for 24 h. GFP levels were measured after 24 h and were normalized to untreated controls. WT cells that did not harbor any promoter reporter were used as a background control. Selected upregulated genes were deleted in the WT cells individually or in combination and mutagenesis assay was performed with 16-min UV exposure. For the second part of the screening, the *E. coli* single deletion mutant library was used, and mutagenesis assay was performed in high-throughput manner using UV radiation for 16 min, followed by 24 h recovery and then plating on rifampicin (RIF)-agar plate to enumerate RIF resistant colony counts. Potential candidate genes from this screening were deleted in MG1655 WT background and mutagenesis assay was performed again in test tubes to confirm the outcomes of the high-throughput screenings.

88f Generation of Species Cross-Reactive CD98hc Antibodies Using Directed Evolution

Neelan Sivaneri, Wenhao Wu, Ece Akin, Peter Tessier

CD98hc, the heavy chain of the large neutral amino acid transporter, has been identified as an attractive target for the transport of therapeutics across the blood-brain barrier (BBB). However, chemical and structural differences between mouse and human CD98hc prevent species cross-reactive binding of anti-CD98hc antibodies, requiring CD98hc transgenic rodents for preclinical studies, which is not possible in some cases and generally impedes clinical translation. In this work, we identify CD98hc epitopes compatible with species cross-reactive binding and use directed evolution to introduce mouse cross-reactive binding in an existing anti-human CD98hc antibody. Random mutations were introduced into one of the complementarity-determining regions (CDRs) of an existing human CD98hc antibody, specifically heavy chain CDR3 (HCDR3). This panel of mutants was sorted for simultaneous binding to both mouse and human CD98hc. A lead clone with micromolar affinity to mouse and human CD98hc was identified through this process, which was subsequently affinity matured by mutagenizing additional CDRs (HCDR1 and HCDR2) and screening for variants with increased affinity using yeast-surface display. This resulted in the isolation of antibodies with nanomolar affinities to both mouse and human CD98hc. These species cross-reactive antibodies are currently being evaluated as bispecific CD98hc antibodies for brain shuttling applications.

88g Integrating Amorphous Solid Dispersions (ASDs) in a PBPK Software

Pedro Monteiro, Pedro Valente

Physiologically based pharmacokinetic (PBPK) software for oral drug delivery typically revolves around the intricate interplay between drug permeation, distribution, and clearance dynamics. While dissolution kinetics for crystalline active pharmaceutical ingredients (APIs) are well understood and can be captured by a reduced set of parameters, such as dissolution rate and the challenge amplifies significantly for amorphous solid dispersions (ASDs). Unlike crystalline forms, ASDs do not face limitations imposed by crystalline solubility, leading to supersaturation in the gastrointestinal tract and is often followed by a subsequent sudden or progressive crystallization of the API ("spring-and-parachute" effect), causing a reduction in the dissolved API amount. Most PBPK software have challenges handling this phenomenon accurately, mis predicting API fraction absorbed.

The distinctive dissolution behavior among different ASDs poses challenges in formulating for optimal *in vivo* performance. PBPK modeling becomes crucial to avoid inaccurate decisions based solely on dissolution metrics like area under the curve or maximum concentration. Permeation affects ASD dissolution profiles by reducing API concentration in the bulk intestinal medium and reducing the driving force for nucleation and crystallization. Consequently, *in vitro* dissolution profiles may differ from *in vivo* dissolution and are not enough to rank and select optimal formulations. Thus, PBPK software for ASDs must grasp these complexities for an accurate formulation ranking *in vivo*.

Following up on this need, Open Systems Pharmacology (OPS) software suite [1], an open-source multi-compartment whole-body physiologically based pharmacokinetic model, composed of PK-Sim and Mobi tools, was used as a starting point. This software allows the customization of individual models (e.g., dissolution) and seamless integration with other software and programming languages (using Mobi). This groundwork enables comprehensive modeling of *in vivo* behavior, covering permeation, clearance, distribution, and excretion models for diverse subjects, with the sole absence of an accurate dissolution model for ASD's.

To circumvent this limitation, a model was developed to captures the specific dissolution characteristics of ASDs using a diffusion-based approach to avoid limiting API release based on crystalline solubility. This strategy required considering two distinct population balances: one accounting for initial ASD powder particle size distribution and its decrease as API and excipients dissolve into the medium, and another for crystallized particle populations formed due to supersaturation-induced nucleation and growth. By resolving the intra particle diffusion of each component the impact of the drug load and excipients solubility can also be captured.

Integrating both models was simplified by leveraging Mobi's compatibility with the R programming language, eliminating the need for software recompilation or significant limitations in dissolution model implementation. A coupled approach was adopted, with the dissolution profile pre-calibrated for dissolution, nucleation, and crystallization parameters based on *in vitro* data being iteratively updated based on permeated drug levels, while PK-Sim handled permeation and clearance. The model bypasses additional emphasis on drug distribution among remaining body compartments, as the ASD formulation's primary effects are already addressed, enabling comprehensive understanding of formulation ranking (by exposure) and exposure.

[1] Lippert J, Burghaus R, Edginton A, et al. Open Systems Pharmacology Community-An Open Access, Open Source, Open Science Approach to Modeling and Simulation in Pharmaceutical Sciences. CPT Pharmacometrics Syst Pharmacol. 2019;8(12):878-882. doi:10.1002/psp4.12473

609a Engineering Filamentous Phages to Deliver Antibacterial Cargo

Yanxi Yang, Shelby Vexler, Irene Chen

Increasing antimicrobial resistance in bacterial pathogens has prompted a renewed interest in alternative approaches such as phage therapy. We are developing a platform for controlled phage therapy using engineered filamentous phages to delivery nanomaterials or antibiotics. The use of a model phage, M13, allows genetic engineering for binding specificity and biosafety. Chemical conjugation of nanoparticle or antibiotic cargo determines the localized effect. For example, phages carrying gold nanorods exhibit bacterial binding and controlled photothermal activation for efficient cell-killing. Studies in our group have validated the safety and efficacy of this approach using in vitro studies as well as a mouse model of *Pseudomonas aeruginosa* wound infection. In particular, wounds infected with a bacterial strain resistant to last-line antibiotics could be successfully treated using the phage-gold nanorod conjugates. In another approach, we engineer M13 to enhance delivery and specificity of antimicrobials that have toxic side effects, such as polymyxin B. In one case, the phage-drug conjugates lowered the required antibiotic concentration by almost two orders of magnitude. These findings point toward potential therapeutic strategies to treat bacterial infections.

609b De-Risking the Implementation of Single Use Reactors for Antibody-Drug Conjugation Reactions.

Catherine Barton

Antibody drug conjugates (ADCs) represent a unique biologics modality wherein a monoclonal antibody is conjugated to a small molecule drug-linker. This modality presents unique manufacturing challenges, including potent solids handling, interaction of large and small molecules, and bioburden risk due to aqueous solutions.

As the demand for Antibody Drug Conjugates (ADC's) increases, the implementation of single-use equipment is vital to optimizing plant productivity. However, concerns regarding inefficient mixing, poor heat transfer, higher sheer rates, oxygen ingress, and extractables have historically prevented the use of single use mixers (SUMs) as conjugation reactors. Significant work has been done to de-risk the scale-up from the lab setup to the production scale SUMs and the transition from the conventional reactors. Both modeling approaches and engineering studies have been applied to provide confidence that the Mobius PowerMix SUM will perform adequately in conjugation processes. This presentation will summarize the efforts in enabling the Mobius PowerMix in clinical ADC programs, sharing the detailed engineering assessment completed.

609c Process Development and Scale-up of ADC Conjugation Processes

Hursh Sureka, Jiajie Huo, Shelby McCowen, Donggeon Nam, Nicolas Cuniere, Selin Aytar, Michael Smith

Antibody-drug conjugates (ADCs) are a class of pharmaceuticals that use an antibody to selectively deliver a cytotoxic drug to cancer cells. Process development for the purification of these modalities via Tangential Flow Filtration (TFF) typically requires screening of several conditions at gram scales to develop robust processes. Via the establishment and use of platform methods and high-throughput screenings, the development of these processes can be accelerated, enabling rapid scale-up while maintaining key product quality attributes. Modeling of both total and individual impurity removal enables deep understanding of the TFF process and is scalable from the lab to the plant. This presentation will highlight these methodologies to enable strategic process development for ADC purification via TFF. Model studies will be presented where the approach was successfully used for process development and subsequent scale-up for non-clinical deliveries.

609d Mitochondria-Targeted Gene Therapy in Combination with Chemotherapy to Treat Triple-Negative Breast Cancers

Tanvi Varadkar, Jiashuai Zhang, Zhuoxin Zhou, Alyssa Fairgrieve, Lufang Zhou, X. Margaret Liu

Out of 1.7 million breast cancers, Triple-negative breast cancer (TNBC) is the most aggressive and metastatic subtype. Conventional chemotherapeutics administered to TNBC patients induce synthetic lethality and have a response rate less than 20%. We have in turn developed a cancer specific luminoptogenetics based gene therapy that causes severe mitochondrial dysfunction thereby causing cancer cell death. We also developed the gene delivery vehicle, i.e. anti-CD276 mAb-tagged exosome-associated adeno-associated virus (mAb-Exo-AAV), to target and kill TNBC cells *in vivo*. The objective of this study was to fully evaluate the novel gene therapy as monotherapy, in combination with a chemotherapy, i.e. poly (ADP-Ribose) polymerase inhibitor (PARPi), and in combination with antibody drug conjugate (ADC), to treat TNBCs. The anti-TNBC efficacy was assessed in multiple animal models, including primary xenograft, metastatic, and PDX mice models. Our results showed that the CD276 mAb-Exo-AAV specifically targeted TNBC cells *in vivo*; the delivered gene therapy effectively depolarized cancer mitochondria; and TNBC tumour burden was eliminated with severe cancer cell death and collapse of tumour microenvironment. The metastasis was significantly reduced by our gene therapy in distant metastatic models. The long-term treatment using combined therapies showed that TNBC recurrence was blocked. Taken together, an innovative gene therapy and delivery vehicle were developed successfully to treat TNBCs. In our future study, we will further investigate the mechanisms of metastasis inhibition and evaluate its toxicology.

609e Anti-Tumor Immune Activation Using Antibody-Drug Conjugates

Hyeyoung Kwon, Greg Thurber

Antibody-drug conjugates (ADCs) form a class of targeted therapeutics that combine an antibody backbone with cytotoxic payloads conjugated by a chemical linker. They can exhibit multiple mechanisms of action (MoA) including Fc-mediated effector function, receptor signal blockade, and direct payload delivery. However, the relative contribution of each component is not fully understood, particularly in the context of an intact immune system. While ADC development typically focuses on payload delivery, studies suggest that ADCs can modulate and synergize with an immune response. To investigate anti-HER2 ADCs for these effects, we 1) tested trastuzumab-based ADCs in immunodeficient and immunocompetent mice using a HER2 expressing syngeneic mouse model then 2) compared the efficacy between LALAPG-, WT-, and Fc-enhanced-trastuzumab to understand the role of Fc-effector functions. We also 3) tested our novel high-avidity, low affinity (HALA) antibodies as 'carrier' antibodies for overcoming heterogeneous distribution challenges in solid tumors in addition to maximizing the antitumor effects.

First, mouse mammary carcinoma E0771 cells transduced with human HER2 lentiviral particles were used to form HER2+ breast cancer (E0771-HER2) tumors either in immunodeficient (nu/nu) or immunocompetent HER2+ transgenic C57BL/6 female mice to create a syngeneic mouse model that expresses human receptors. Many ADC payloads have lower potency in mouse cells than human cells, so payloads were screened for efficacy in mouse cancer cells *in vitro*. PBD and PNU-159682 were identified as having suitable potency. Next, E0771-HER2 cells were injected in the hind limb and mammary fat pad for nude mice and HER2 transgenic mice, respectively. Once the tumor volume reached 250 mm^3, 1 mg/kg of trastuzumab-PBD (T-PBD) or trastuzumab-PNU was administered through tail vein injection in nude mice to study the efficacy of payload-mediated cell death. To compare the impact of immune effects during ADC treatment, we tested the same dose of T-PBD in the immunocompetent HER2+ syngeneic model. To quantify the impact of Fc-effector functions, combinations of LALAPG-T-PBD (no Fc effector function) and WT-T-PBD (normal Fc effects) were administered with or without an enhanced-Fc HALA antibody (eFc-HALA; enhanced ADCC and ADCP effects) at 1:3 (ADC:carrier) ratio. The tumor volumes and weight of the mice were monitored every other day after the administration. Cured mice were rechallenged with both WT E0771 or E0771-HER2 to test for immunological memory.

With E0771-HER2 positive cells, DNA damaging agents (e.g., PBD, PNU-159682, Calicheamicin D) showed high potency with picomolar IC50s *in vitro*. However, FDA-approved Kadcyla and Enhertu had IC50 values greater than 100-fold higher, indicating their potency is low in this mouse cell line. We then investigated the payload efficacy *in vivo* by administering 1 mg/kg of T-PBD and T-PNU to immunodeficient nude mice. T-PBD resulted in moderate growth inhibition but not tumor reduction in immunodeficient mice. Interestingly, T-PNU given at the same dose showed a significant improvement with tumor reduction after initial tumor growth. This demonstrated that the payload efficacy in vivo, even if they have similar IC50 values in vitro, may differ significantly. Studies in tumor spheroids indicated that the bystander killing of PNU was much more efficient than PBD. To test the efficacy in the presence of an intact immune system, the same dose of T-PBD was given to transgenic mice bearing E0771-HER2 tumors. Compared to the efficacy in the immunodeficient mice, T-PBD in the immunocompetent model resulted in tumor regression with 4 out of 10 complete responses, indicating a synergistic effect of the ADC with the immune system. We then investigated the impact of the Fc-effector function in antitumor activity with LALAPG-T-PBD. LALAPG-T-PBD lacks Fc-gamma receptor binding, where the ADC only delivers payload to the tumor. Without Fc receptor binding, LALAPG-T-PBD showed only one complete response. The 1 mg/kg dose is not sufficient to saturate the tumor in this moderately expressing HER2+ cell line, resulting in heterogeneous distribution. In order to overcome this delivery limitation of ADCs in solid tumor and enhance ADCC/ADCP responses, we co-administered an eFc-HALA carrier dose with WT T-PBD. HALA antibodies are engineered to selectively compete with ADCs on high-expressing cells while avoiding competition on low-expressing cells, thus, allowing ADCs to penetrate deeper into the tissue while maintaining sufficient uptake on all cells. We further engineered the Fc-binding of HALA antibody to increase the affinity to FcγRII and FcγRIII which are known to be responsible for ADCC and ADCP. While a control cohort of eFc-HALA only had one partial responder out of 10, T-PBD with eFc-HALA had 9 out of 11 partial and complete responses, resulting in the highest overall response rate. This result indicates that the co-administration of eFc-HALA antibody with an ADC can improve the survival rate via synergy between payload cell killing and Fc effector function in antitumor activities.

In summary, we utilized a syngeneic mouse model with E0771-HER2 tumors in immunocompetent HER2 transgenic mice to test the efficacy of trastuzumab-based ADCs in an immunocompetent setting. We compared the efficacy of T-PBD in nude mice and transgenic mice and observed greater efficacy with the presence of the immune system. While WT T-PBD showed a range of responses (no, partial, and complete responses), Fc-null T-PBD reduced tumor size but resulted in only one complete responder. Using the syngeneic model, we further tested ADCs with eFc-HALA carrier dose which increases the ADC tumor penetration and ADCC and ADCP responses. T-PBD with eFc-HALA co-administration cohort showed the highest ORR, indicating that the payload and Fc effector function have synergistic effects in tumor elimination. Our findings have shown that we were further able to maximize the efficacy of ADCs by designing their payloads to recruit and activate immune cells in addition to engineering their antibody backbone to increase affinity to Fc-receptors on NK cells and macrophages.

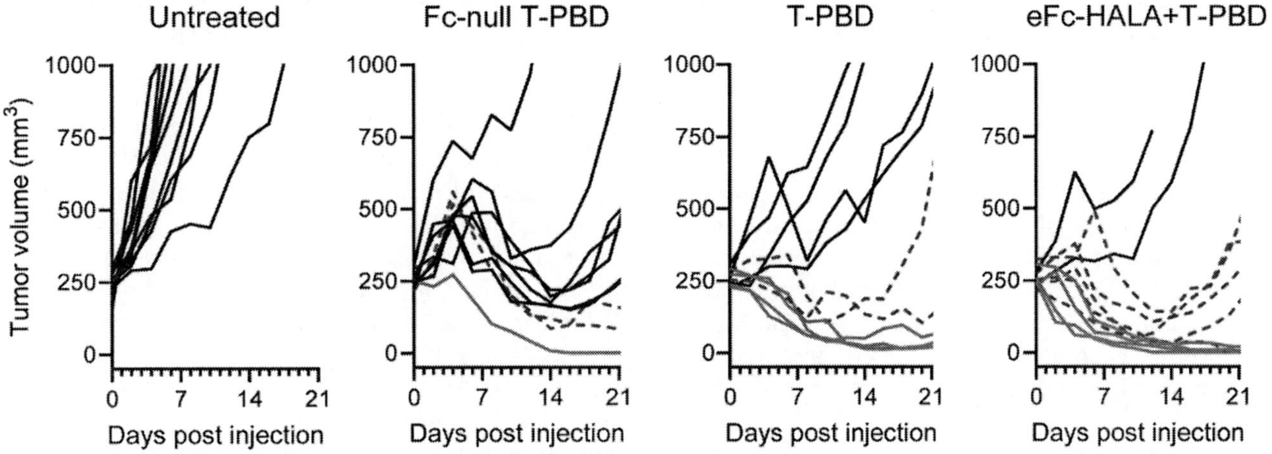

	Overall Response Rate (%)	Partial Response	Complete Response
Control	0/10 (0%)	0	0
Fc-null T-PBD	3/10 (30%)	2	1
T-PBD	6/10 (60%)	2	4
eFc-HALA+T-PBD	9/11 (82%)	5	4

Partial Response:
> 50% tumor reduction

Complete Response:
No deteactable tumor

609f Microfluidic Engineering of pDNA Nanogels Using a Co-Axial Flow Reactor

Suneha Patil, Zoe Whiteley, Esther Osarfo-Mensah, Duncan Craig, Stefan Guldin, Asterios Gavriilidis

Gene therapy requires vectors that enable the safe transport of genetic material across cell membranes. Nanogels (NGs) posses multifunctional characteristics viz. sizes <200 nm which enable them to enter the cell via endocytosis and high drug loading and encapsulation capability for a wide range of bioactive molecules such as pDNA, mRNA, oligonucleotides. Their biocompatibility, safety as carriers, and diverse drug release properties (stimuli-responsiveness, controlled response at the target site), render them an excellent model to be used as carriers for gene therapy. In this work, a coaxial flow (CFR) reactor is used to synthesize Carboxymethyl chitosan grafted Polyethylenimine nanogels (CMC-bPEI-pDNA NGs) as carriers for pDNA delivery, due to its laminar flow mixing, suitable for shear sensitive pDNA, ease of operation, and a simple and economic reactor assembly. The CFR is made up of an inner capillary (Inner diameter 0.14 mm) inserted into a larger capillary (Inner diameter 1.12 mm) to form a core flow surrounded by a sheath flow. Mixing occurs at the point where the core and sheath streams meet (outlet of the inner capillary), and the reaction proceeds across the length of the reactor. For polymeric NGs, the cationic polyplexes formed by the polymer complex and pDNA, not only condense the pDNA into smaller particles but are also expected to adhere to the negatively charged cell membrane via electrostatic interactions, thus overall facilitating cellular uptake.

We used the CFR to engineer cross-linked cationic nanogels of diameter <200 nm, with a polydispersity index (PDI) <0.3, encapsulation efficiency (EE) >90% and a high transfection efficiency (>80%). We performed a systematic parametric optimisation including variation of residence time, flow ratio, NP ratio (number of moles of nitrogen in polymer to moles of phosphate in pDNA and cross-linker), cross-linker concentration and pDNA loading. The reagents, polymer solution was added through the core capillary, while the pDNA and cross-linker through the sheath capillary of the CFR, at a particular flow ratio - FR (ratio of core flowrate to the sheath flowrate). The CFR decreased the reaction time for complexation of nanogels to as low as 7 s. All NGs prepared with NP ratios from 0.2 to 48 had sizes below 200 nm (measured using Dynamic Light Scattering) and in the range for endocytic uptakes. However, these NGs were negatively charged for NP < 5. It is interesting to note that the surface charge on NGs can be manipulated in a CFR from cationic to anionic at higher flow ratios (core/sheath) FR>2 or when the location of the reagent streams were interchanged. Hence, cationic monodisperse nanogels with diameter <200 nm and %EE>85% were produced. This tuneability of the overall charge is advantageous as it can avoid the requirement of an additional coating on the nanogels performed to render them cationic for systemic administration. We also found that the encapsulation efficiency can be increased to >99.5% by increasing the cross-linker concentration. Furthermore, we could load pDNA as high as 75 ng/μL and still attain highly monodisperse NG sizes: ~165 nm, PDI < 0.1, EE>99%. The final optimized formulation of nanogels was then scaled-up using a larger CFR and increasing the throughput tenfold. Overall, these NGs successfully deliver the payload to HEK293T cells, similar to the commercially available non-viral vector Lipofectamine 3k after 72 hours.

609g Using Mathematical Models to Accelerate Peptide Synthesis

Shekhar Viswanath, Jingyao Wang, Shankar Vaidyaraman, Bhoja Kandela

This talk will provide an overview of various mathematical models developed to accelerate peptide process development. We will start with a mathematical foundation for optimal fragmentation to balance estimated cost and throughput. Next, we will pivot to provide an overview of mathematical models built to accelerate commercial development of solid phase peptide synthesis as well as reverse phase chromatography unit operations.

423a State Estimation for Manufacturing of mRNA-Based Vaccines and Therapeutics: A Lyophilization Case Study

Prakitr Srisuma, George Barbastathis, Richard Braatz

Abstract:

Lyophilization (aka freeze drying) is a process used to increase the stability of biotherapeutics, e.g., mRNA vaccines, in biopharmaceutical manufacturing [1], allowing for their storage at higher temperature while preserving their functionality [2, 3]. This advance could play an important role in future mRNA-based therapeutics, including the enabling of vaccine distribution in regions where a cold supply chain is lacking.

Lyophilization consists of three main stages, namely (1) freezing, (2) primary drying, and (3) secondary drying, respectively. The final stage, secondary drying, entails bound water removal, in which the accurate prediction and monitoring of bound water concentration are the key to ensuring the quality of the lyophilized product [4, 5]. Various measurement and monitoring techniques have been proposed and studied [6]. One of the most common techniques is the Karl Fischer titration, which requires sampling of the vial for offline measurements [6–8]. To avoid process interruption, some online or non-invasive techniques such as near-infrared (NIR) spectroscopy [9] and tunable diode laser absorption spectroscopy (TDLAS) [10] have been proposed. Detailed discussion of tools for the monitoring of secondary drying can be found in Ref. [6].

Instead of direct measurement, a state observer (aka state estimator, observer, estimator) can be used to estimate states that are not measured [11]; the process is known as state estimation. A state observer combines the available measurements and mechanistic understanding of a system (mechanistic model) to optimally estimate the unmeasured states. A well-designed observer can replace expensive and complicated sensors in the system, reducing the total cost and complexity of operation. In the context of lyophilization, state observers have been studied and applied to the primary drying step, which aims at estimating the temperature, interface position (amount of ice), and relevant parameters such as the heat transfer coefficient [12–19]. Observer design for primary drying is straightforward as it mainly concerns heat transfer associated with sublimation and, in many cases, an observer is not even needed as heat transfer-related quantities, e.g., temperature, can be measured easily. In secondary during, heat transfer and desorption dynamics are coupled, making the observer design more challenging and valuable. Currently, applications of state estimation to secondary drying are very limited. To our knowledge, the only literature that proposed a state estimation-like strategy for secondary drying is Ref. [8]; the technique is referred to as a soft sensor which requires the measurement of the desorption flux for estimating the residual moisture. The procedure in Ref. [8] does not exploit the mathematical structure of a state observer; the key idea is to iteratively solve the optimization to find the moisture content that matches the measured desorption flux. This technique also requires additional equipment specifically for measuring the desorption flux.

In this work, a novel technique is introduced for real-time estimation of bound water concentration during desorption, which is applied to secondary drying in lyophilization. A state observer is mathematically formulated to estimate the concentration of bound water given the mechanistic model and temperature measurement. A practical observer design strategy is described and demonstrated in detail. Our observer can accurately estimate the concentration for various desorption dynamics, noisy data, and real experiments. Nearly all the case studies presented in this work are achieved by a single observer design, indicating high robustness of the observer.

In terms of practicality and implementation, our observer is designed such that it can be simulated in real time, with the computation time of less than a second on a normal laptop. Besides, no concentration measurement is required; the only measurement data required is temperature. Temperature measurement is straightforward and very common in every step of lyophilization [6, 20], whereas accurate bound water measurement is not trivial and involves complex equipment and procedures [21]. Therefore, our technique allows for the simplest setup and operation compared to any other methods, eliminating any additional sensor and complexity. The framework can also be easily and systematically extended to other desorption-based processes.

Acknowledgements:

This research was supported by the U.S. Food and Drug Administration under the FDA BAA-22-00123 program, Award Number 75F40122C00200.

References

[1] A. Liapis, R. Bruttini, A theory for the primary and secondary drying stages of the freeze-drying of pharmaceutical crystalline and amorphous solutes: Comparison between experimental data and theory, Separations Technology 4 (3) (1994) 144–155. doi:10.1016/0956-9618(94)80017-0.

[2] H. Muramatsu, K. Lam, C. Bajusz, D. Laczk´o, K. Karik´o, P. Schreiner, A. Martin, P. Lutwyche, J. Heyes, N. Pardi, Lyophilization provides long-term stability for a lipid nanoparticle-formulated, nucleoside-modified mRNA vaccine, Mol. Ther. 30 (5) (2022) 1941–1951. doi:10.1016/j.ymthe.2022.02.001.

[3] S. Meulewaeter, G. NuFytten, M. H. Cheng, S. C. De Smedt, P. R. Cullis, T. De Beer, I. Lentacker, R. Verbeke, Continuous freeze-drying of messenger RNA lipid nanoparticles enables storage at higher temperatures, J. Control. Release 357 (2023) 149–160. doi: 10.1016/j.jconrel.2023.03.039.

[4] S. A. Velardi, A. A. Barresi, Development of simplified models for the freeze-drying process and investigation of the optimal operating conditions, Chemical Engineering Research and Design 86 (1) (2008) 9–22. doi:10.1016/j.cherd.2007.10.007.

[5] J. H. Gitter, R. Geidobler, I. Presser, G. Winter, Microwave-assisted freeze-drying of monoclonal antibodies: Product quality aspects and storage stability, Pharmaceutics 11 (12) (2019). doi:10.3390/pharmaceutics11120674. [4] S. A. Velardi, A. A. Barresi, Development of simplified models for the freeze-drying process and investigation of the optimal operating conditions, Chemical Engineering Research and Design 86 (1) (2008) 9–22. doi:10.1016/j.cherd.2007.10.007.

[6] D. Fissore, R. Pisano, A. A. Barresi, Process analytical technology for monitoring pharmaceuticals freeze-drying – A comprehensive review, Dry. Technol. 36 (15) (2018) 1839–1865.doi:10.1080/07373937.2018.1440590. [4] S. A. Velardi, A. A. Barresi, Development of simplified models for the freeze-drying process and investigation of the optimal operating conditions, Chemical Engineering Research and Design 86 (1) (2008) 9–22. doi:10.1016/j.cherd.2007.10.007.

[7] M. J. Pikal, W. J. Mascarenhas, H. U. Akay, S. Cardon, C. Bhugra, F. Jameel, S. Rambhatla, The nonsteady state modeling of freeze drying: In-process product temperature and moisture content mapping and pharmaceutical product quality applications, Pharmaceutical Development and Technology 10 (1) (2005) 17–32. doi:10.1081/PDT-35869.

[8] D. Fissore, R. Pisano, A. A. Barresi, Monitoring of the secondary drying in freeze-drying of pharmaceuticals, Journal of Pharmaceutical Sciences 100 (2) (2011) 732–742. doi:10.1002/jps.22311.

[9] C. Ikeda, G. Zhou, Y.-C. Lee, G. Chouzouri, L. Howell, B. Marshall, L. Bras, Application of online NIR spectroscopy to enhance process understanding and enable in-process control testing of secondary drying process for a spray-dried solid dispersion intermediate, Journal of Pharmaceutical Sciences 111 (9) (2022) 2540–2551. doi:10.1016/j.xphs.2022.04.009.

[10] S S. C. Schneid, H. Gieseler, W. J. Kessler, S. A. Luthra, M. J. Pikal, Optimization of the secondary drying step in freeze drying using TDLAS technology, AAPS PharmSciTech 12 (1) (2011) 379–387. doi:10.1208/s12249-011-9600-7.

[11] D. Luenberger, An introduction to observers, IEEE Transactions on Automatic Control 16 (6) (1971) 596–602. doi:10.1109/TAC.1971.1099826.

[12] S. A. Velardi, H. Hammouri, A. A. Barresi, In-line monitoring of the primary drying phase of the freeze-drying process in vial by means of a kalman filter based observer, Chemical Engineering Research and Design 87 (10) (2009) 1409–1419. doi:10.1016/j.cherd.2009.03.011.

[13] H. H. Salvatore A. Velardi, A. A. Barresi, Development of a high gain observer for in-line monitoring of sublimation in vial freeze drying, Drying Technology 28 (2) (2010) 256–268.

doi:10.1080/07373930903530204.

[14] S. Bosca, D. Fissore, Design and validation of an innovative soft-sensor for pharmaceuticals

freeze-drying monitoring, Chemical Engineering Science 66 (21) (2011) 5127–5136. doi:

10.1016/j.ces.2011.07.008.

[15] Bosca, A. A. Barresi, D. Fissore, Use of a soft sensor for the fast estimation of dried cake resistance during a freeze-drying cycle, International Journal of Pharmaceutics 451 (1) (2013) 23–33. doi:10.1016/j.ijpharm.2013.04.046.

[16] S. C. Elena Niculina Drăgoi, D. Fissore, On the use of artificial neural networks to monitor a pharmaceutical freeze-drying process, Drying Technology 31 (1) (2013) 72–81. doi:10.1080/07373937.2012.718308.

[17] A. A. B. Serena Bosca, D. Fissore, Design of a robust soft-sensor to monitor in-line a freeze-drying process, Drying Technology 33 (9) (2015) 1039–1050. doi:10.1080/07373937.2014.982250.

[18] A. A. B. Serena Bosca, D. Fissore, On the robustness of the soft sensors used to monitor a vial freeze-drying process, Drying Technology 35 (9) (2017) 1085–1097. doi:10.1080/07373937.2016.1243553.

[19] D. Fissore, R. Pisano, A. A. Barresi, On the use of temperature measurement to monitor a freeze-drying process for pharmaceuticals, in: IEEE International Instrumentation and Measurement Technology Conference, 2017, pp. 1–6. doi:10.1109/I2MTC.2017.7969890.

[20] M. Harguindeguy, L. Stratta, D. Fissore, R. Pisano, Combining mathematical modeling and thermal infrared data in the freezing of pharmaceutical liquid formulations, Industrial & Engineering Chemistry Research 61 (12) (2022) 4379–4389. doi:10.1021/acs.iecr.1c04595.

[21] M. U. H. Joardder, M. Mourshed, M. Hasan Masud, Bound Water Measurement Techniques, Springer International Publishing, Cham, 2019, pp. 47–82. doi:10.1007/978-3-319 99888-6_4.

423b Advancing Solubility Prediction Using an AI-Driven Tool for Resource Optimization in Pharmaceutical Research

Laura Pereira Diaz, Adrian Ramirez Galilea, Andrew Kukor, Neda Nazemifard, Mark Henson

In recent years, the application of data-driven models has been adopted by various research and development fields such as materials science, chemistry, and drug discovery and development. Machine learning techniques are now commonly used to predict physicochemical properties and process parameters. In materials science, the availability of large amounts of data has led to the development of advanced tools for materials informatics. While mechanistic models have been used to predict solubility values, machine learning methods have also been gaining attention. Deep learning techniques, such as convolutional neural networks, have emerged as a promising approach for capturing complex structure-property relationships for property prediction. Computer vision techniques have also been used to predict solubility by analyzing images of the solute and solvent [1]. By leveraging these data-driven approaches, scientists and engineers can develop more accurate and efficient models for predicting solubility, which can have important applications in fields such as drug discovery and material science.

Predicting solubility of complex pharmaceutical compounds is crucial at several stages of active pharmaceutical ingredient process development. Through the lens of green chemistry, solvent selection has become increasingly important in process design and optimization, aiming to reduce the use of the most hazardous, environmentally damaging solvents [2]. Although traditionally this challenge has been addressed by experimental solubility measurements [3], these are time- and material-consuming, limiting the opportunity of solvent screening. Hence, there is a need for computational models that can predict compound solubility efficiently, not only accelerating but optimizing the early stages of pharmaceutical development.

Hereby, a novel approach for predicting the solubility of chemical compounds is presented. The proposed method is based on a hybrid neural network architecture that combines deep learning, computer vision, and experimental input data (Figure 1). To the best of the authors' knowledge, there is no other model in the literature that uses such a hybrid architecture for solubility prediction. The proposed model leverages deep learning to extract features from images and experimental data, and of computer vision to analyze and interpret these features. The model is trained on a large dataset of over 240 systems and associated solubility and turbidity measurements and is optimized using continuous refinement and optimization techniques. The authors demonstrate that the proposed model achieves remarkable accuracy in predicting solubility (98%), outperforming other state-of-the-art models. The model has been validated through a robust cross-validation test.

The proposed model has the potential to significantly improve the efficiency of solubility screening in early-stage research and development, and to reduce the experimental workload and resource consumption in the chemical and pharmaceutical industries. A data-driven approach for solubility screening can significantly reduce the time and resources required for experimental solvent screening. Furthermore, data-driven models have the potential to outperform traditional mechanistic models [4]. This study presents a promising direction for future research in the field of solubility prediction and highlights the potential of hybrid neural network architectures for solving complex problems in chemistry and related fields.

References

1. Singh, V., et al., *Recent trends in computational tools and data-driven modeling for advanced materials.* Materials Advances, 2022. **3**(10): p. 4069-4087.
2. Byrne, F.P., et al., *Tools and techniques for solvent selection: green solvent selection guides.* Sustainable Chemical Processes, 2016. **4**: p. 1-24.
3. Könczöl, Á. and G. Dargó, *Brief overview of solubility methods: Recent trends in equilibrium solubility measurement and predictive models.* Drug Discovery Today: Technologies, 2018. **27**: p. 3-10.

4. Piccard, P.-J., et al., *Organic Solvent Nanofiltration and Data-Driven Approaches.* Separations, 2023. **10**(9): p. 516.

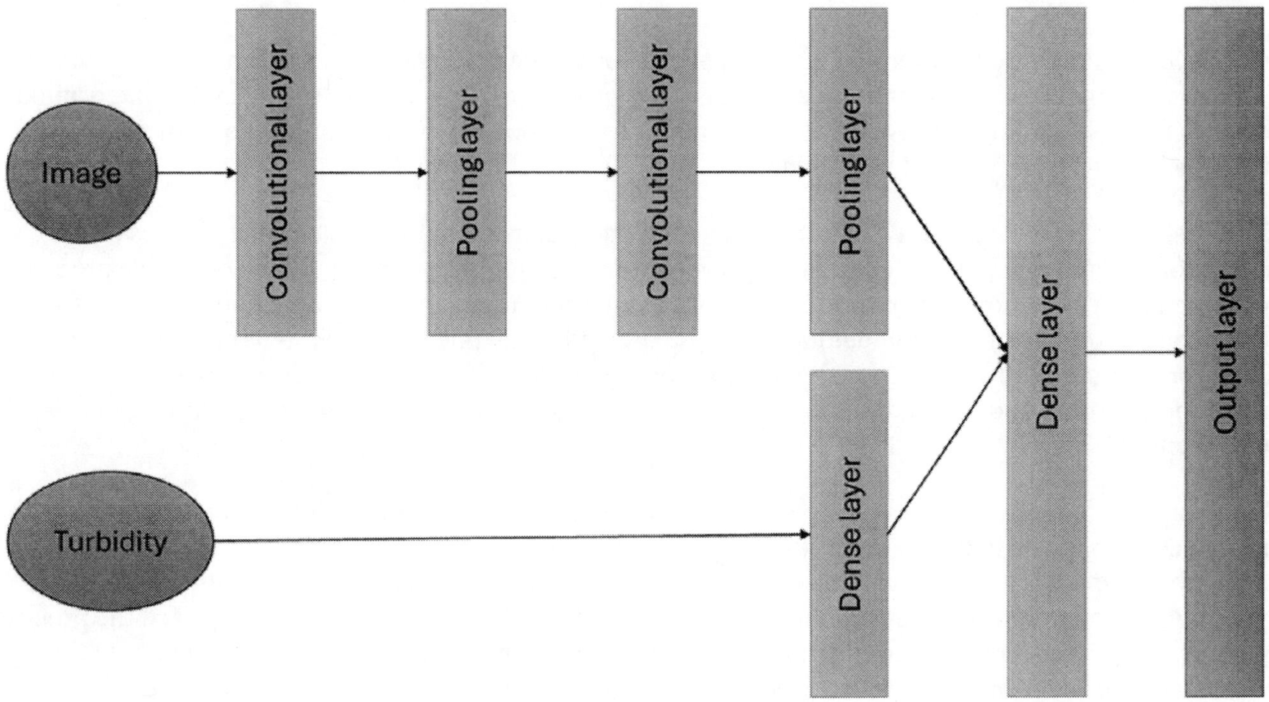

423c Sequential, Simulation-Based Manufacturing Execution System for the Planning of Pharmaceutical Production

Daniel Casas, Salvador García-Muñoz

Fast development timelines for new medicines call for accelerated engineering workflows that enable evaluation of manufacturing options while ensuring product quality, patient safety and uninterrupted supply. This evaluation can involve resource planning and operation forecasting (equipment, personnel, materials) to fulfill the demand of a product in a timely manner.

Different approaches exist to study production and operations planning. Scheduling and resource allocation optimization approaches based on mixed-integer problem formulations are a wide area of research [1]. A common approach to modeling the unit operations in these contributions is that of using coarse models of low computational cost that neglect the phenomena internal to the process. Therefore, forcing the analyst to make assumptions about the distribution of disturbances that may affect the planning and scheduling, even when these variations can have a physical cause associated with them.

Another less common approach is the use of process simulators coupled with planning tools as an upper decision layer. In this case, task durations as well as the internal behavior of the processing units are retrieved after solving material and energy balances. This not only helps with operation planning, but also to process understanding as information of the physics and chemistry of the process becomes available. An important feature of the said simulators is the ability to deal with a combination of continuous and batch/semi-batch operations, as is often encountered in the pharmaceutical industry. To this end, we leverage the numerical library PharmaPy [2] to model the operation (resource consumption, schedule, timings) of a chemical manufacturing plant.

In this work, we propose a methodology to simulate the operation and scheduling of an entire chemical plant given a series of models for each unit operation involved, and the information about the resources available and those necessary for each unit operation to be able to run. In our approach, a Manufacturing Execution System (MES) is simulated using a sequential-oriented engine to manage the execution of subsets of unit operation models involved in the flowsheet, as opposed to an equation-oriented approach that solves the entire flowsheet simultaneously. To this end, the transfer of material across the different material reservoirs and operations in the flowsheet is managed by the simulated MES by following pre-established logical rules about the operation. This approach enables us to study the effect of the variations of process consumables (e.g. quality of raw materials) and available resources (number of intermediate storage units) onto the overall operating schedule of a facility.

Our methodology is demonstrated by analyzing multiple process configurations of a hypothetical flowsheet for the manufacture of an active pharmaceutical ingredient (API), where a continuous synthesis step is followed by batch separation and purification. The fictitious process includes multiple batch processing units and storage tanks, for which the MES decides task assignment and process timing/sequencing. Of particular interest, scenarios of increased throughput for a fixed plant configuration, as well as alternatives with less or more processing or storage units are analyzed and compared based on makespan and mean throughput. While in this contribution we utilize a concocted scenario to illustrate the features of the methodology, the authors have applied this technique to aid the study and optimization of a real manufacturing train.

[1] Harjunkoski, I., et al., *Computers & Chemical Engineering*, 2014. **62**: p. 161-193.

[2] Casas-Orozco, D., et al., *Computers & Chemical Engineering*, 2021. **153**.

423d Digital Real-Time Release Testing in Continuous Tablet Manufacturing

Selma Celikovic, Jakob Rehrl, Ruben Martins Fraga, Martin Steinberger, Martin Horn, Johannes Khinast, Stephan Sacher, Andreas Kottlan

In traditional pharmaceutical manufacturing, the quality is assured via in-process controls and end-product testing. After approval of samples in the quality control laboratory, the batch is released to the next process step or at the end of the line for the patient. No real-time information is required for this quality by testing (QbT) approach. The implementation of quality by design (QbD) and the application of process analytical technology (PAT) have led to a paradigm change. PAT is an enabler for the evaluation of the material quality and for control actions in real-time, which are essential for continuous manufacturing. Data acquisition during the process is also required for real-time release testing (RTRT). RTRT strategies have been developed for a few unit operations and combinations thereof. However, in continuous manufacturing lines, a huge amount of data is available in addition to PAT sensors, which comes from univariate measurements and from the equipment. Residence time distribution (RTD) models can be used to match the information with the material.

In this study, for the first time comprehensive data from a fully integrated continuous manufacturing line is matched with specific amounts of product. The data are acquired from PAT tools, process equipment, and soft sensors and collected in a process control and data management platform. An RTD model of the plant and a material tracking algorithm are used to match the data with the material. At the end of the manufacturing cycle, the complete history of process conditions is digitally available for each specific amount of product. The approach was implemented and demonstrated at a Consigma 25 tableting line.

423e An Automated Crystallization-Filtration-Drying Workstation: Development and Application from the Perspective of Pharmaceutical Industry

Qinglin Su, Tharanga Wijethunga, David Willcox, Carla Luciani

Automated crystallization platforms have been pursued for more than two decades with the progress of process analytical technology (PAT) tools in the pharmaceutical R&D sector [1]. By manipulating solute concentration, supersaturation levels, and other parameters, crystallization mechanism and kinetics can be better understood to control crystal size, shape, morphology [2]. While fruitful achievements in academy and industrial applications have been reported [3], commercially *off-the-shelf* crystallization platforms or workstations are still not readily available or routinely deployed. Furthermore, downstream operations such as filtration and drying of the final crystalline product are often not integrated. Typically, the overall process cause-and-effect relationship in the fishbone diagram are not fully understood [4].

In this work, the development and application of an in-house automated crystallization-filtration-drying workstation for process development and engineering purposes will be discussed. This includes: (1) an automation I/O communication network to integrate the equipment, instruments/sensors, and mathematical models; (2) process monitoring and PAT-based feed-back control strategies for batch crystallization with cooling, antisolvent addition, or combined cooling and antisolvent addition; (3) wet cake filtration characterization; (4) drying kinetics under heated, vacuumed, and/or humidified conditions. Equipment design, automation, human-machine-interface (HMI), and considerations for further development will be discussed.

References

[1] Fujiwara M, Nagy ZK, Chew JW, Braatz RD. First-principles and direct design approaches for the control of pharmaceutical crystallization. *Journal of Process Control*, 2005; 15(5):493-504.

[2] Nagy ZK, Braatz RD. Advances and new directions in crystallization control. *Annu Rev Chem Biomol Eng*. 2012; 3:55-75.

[3] Gao Y, Zhang T, Ma Y, Xue F, Gao Z, Hou B, Gong J. Application of PAT-Based Feedback Control Approaches in Pharmaceutical Crystallization. *Crystals*. 2021; 11(3):221.

[4] Su Q, Ganesh S, Moreno M, Bommireddy Y, Gonzalez M, Reklaitis GV, Nagy ZK. A perspective on Quality-by-Control (QbC) in pharmaceutical continuous manufacturing. *Computers & Chemical Engineering*. 2019; 125: 216-231.

423f Development and Application of a MULTI-Component MODEL with Radially-Dependent Droplet Drying for Pharmaceutical Spray Drying

Patrick Harris, Suela Jonuzaj, Stefan Bellinghausen, Xin Li, Lok Hang Wong, Hassan Abdullahi, Yogesh Patil, John Rose, Christopher L. Burcham

Spray drying is commonly employed in the pharmaceutical industry to form amorphous solid dispersions (ASD) that enhance the dissolution properties of the active pharmaceutical ingredient (API) in the drug product. Effective process design is essential to ensure desired product properties such as particle size and residual solvent content, while maintaining satisfactory physical stability. By combining the use of digital tools with experimentation, the design and scale up of spray drying processes may be improved.

In this work, we detail the development of a multi-component spray drying model in the gPROMS FormulatedProducts (gFP) framework with two distinct modules: 1) A heat and mass balance spray dryer process model, and 2) A droplet drying model with radial diffusion of components. More advanced thermodynamic models within gFP (e.g., NRTL) as well as dynamic vapor sorption (DVS) and differential scanning calorimetry (DSC) data can be used to inform the first module to simulate the predict product physical stability. With information of the liquid diffusion properties of solid components to inform the second module, the radial distribution of components across the particles may be predicted, allowing the model to predict shell locking. The model may be applied to predict a number of measurable product properties (e.g., particle size, solvent content, glass transition temperature) and their sensitivity to process parameters (e.g. drying gas temperature, solution feed rate, etc.), as well as be used to improve the environmental footprint of the process via reduced process mass intensity (PMI). The model may be utilized from early to late stages of pharmaceutical development, based on the level of knowledge of system properties, to improve and accelerate the development of pharmaceutical spray dried amorphous dispersions. Last, we demonstrate the use of this model for a case study of a pharmaceutical spray dried amorphous dispersion to predict the impact of process conditions on residual solvent content, glass transition temperature with a comparison to experimental process data and identify ranges of process parameters which yield desirable product qualities.

423g The Pursuit of Tighter Titer: Model Predictive Control of Monoclonal Antibody Titer Under Disturbances

Zach Hatzenbeller, Anastasia Nikolakopoulou, Yu Luo

Fed-batch bioreactors are particularly useful for processes that require long cultivation times or where it is difficult to predict the nutrient requirements of the cells [1]. By incrementally adding nutrients, one can help ensure high product yield while maintaining near-optimal culture conditions in fed-batch bioreactors. Batch-to-batch variability in end-of-run titer is a challenge in biomanufacturing operations and can result in cost increases.

The dynamic and complex nature of cell culture systems poses a significant challenge due to the potential for unpredictable process variability. Cell growth and metabolism are affected by a wide range of factors, such as nutrient availability, dissolved oxygen levels, pH, temperature, and shear stress [2]. Subsequently, monoclonal antibody product titer can be affected by cell growth and metabolism. Lastly, the metabolic activity of cells changes over time, which makes it difficult to make accurate predictions. The feeding strategy can be optimized based on the specific cell line in addition to culture conditions and therefore it requires a customized approach tailored to the specific requirements of the production process [3]. However, when disturbances act on the process, the operation is no longer progressing under optimal feeding conditions and is expected to deviate from a desired trajectory. This leaves the process susceptible to low productivity or altered product quality.

Process control is a systematic framework designed to maintain a process within a specified setpoint and enhance its robustness against uncertainties. Simple control loops, such as proportional-integral (PI) control loops are readily implemented in cell culture processes to control the bioreactor's temperature, dissolved oxygen, and pH. The idea of implementing advanced process control of productivity and quality attributes in biomanufacturing is increasingly attracting interest from the industry. A cell culture model can be integrated into advanced control methodologies such as model predictive control (MPC) to control titer "tightly" around a specified nominal trajectory by optimizing the cell culture feeding strategy.

In this talk, we discuss MPC strategies for fed-batch mammalian cell culture processes. MPC strategies are investigated for their potential to achieve consistent end-of-run titer in the presence of unexpected process disturbances from initial viable cell count, pH, and temperature. First, we discuss two different model structures and their integration with MPC. Then, we compare empirically the effects of process disturbances on end-of-run titer with and without MPC, showing that MPC reduces process variability and maintains consistent titers.

References

[1] J. Xu *et al.*, Systematic development of temperature shift strategies for Chinese hamster ovary cells based on short duration cultures and kinetic modeling. *MAbs* **11**, 191-204 (2019).

[2] Y. Chen *et al.*, Digital Twins in Pharmaceutical and Biopharmaceutical Manufacturing: A Literature Review. *Processes* **8**, 1088 (2020).

[3] J. C. Yee *et al.*, Advances in process control strategies for mammalian fed-batch cultures. *Current Opinion in Chemical Engineering* **22**, 34-41 (2018).

184b Polysaccharide Charge Reduction For Conjugation Vaccine Modernization

Megan McCarthy, Katherine Phillips, Christopher Brennan

A lower charge ratio of polysaccharide to protein was investigated as part of modernizing the process for a conjugate vaccine. Historical data indicates that the polysaccharide is not fully consumed in the conjugation reaction. By modifying the amount of polysaccharide charged, we optimized reaction efficiency and enabled a cost savings by improving yields. These experiments also led to an improved understanding of reaction kinetics, which had been previously limited by analytical capabilities. Initial experiments explored the limits of the polysaccharide to protein ratio and mapped reaction kinetics. A thirty-two arm design of experiments further investigated these two factors and supported the polysaccharide charge optimization with no impact to product critical quality attributes. Future studies will explore a broader range of polysaccharide charge ratio to further define optimized and robust conjugation.

184c Need Big Particles but Have No Time? Rapid Development and Scale-up of Spray Drying to Control Particle Size

David Goundie, Benjamin Hawkins

API clinical manufacturing of early-phase assets is often fraught with tight timelines, unexpected deviations, and the need for prompt development solutions. The need for flexible, easy-to-perform experiments is critical to maintain schedules and prevent project delays. With that in mind, the team developed a workflow designed to rapidly understand how to control particle size through spray drying using minimal development conditions. This was necessary in order to reprocess amorphous API to the original particle size and supply material in time for clinical trials. Through literature and internal expertise, key parameters were identified and tested via a lab-scale unit before scaling up to the manufacturing spray dryer. With only two days of development time on the manufacturing equipment prior to clinical spray drying, the new workflow was implemented. After changes to the fluid cap size and visual confirmation of acceptable atomization across a range of atomization nitrogen pressures, three small batches were spray dried at minimum, maximum, and standard atomization pressures. By fitting a trend to the data, an atomization flow rate was calculated to achieve the target particle size for the clinical batch. A final test run showed acceptable particle size and residual solvent levels, which enabled clinical processing to start and delivery of amorphous clinical-grade reprocessed API without further project delays.

184d Empirical Calculations for Reliable Batch Blending Operation Scale-up in Pharmaceutical Manufacturing

Kanishka Ghosh, Rajesh Kulkarni, Joshua Hanson, Joel Francis, Maitraye Sen

Batch blending is an important step in pharmaceutical drug product manufacturing for preparing the final formulation blend that goes into the drug product dosage form. A batch process is initially explored and developed at a small (lab) scale and then scaled up to the intended commercial scale. Currently, there is a gap in the availability of rigorous scale-up criteria for batch blending. A commonly used scale-up factor for blending is 'K-value' that is a scale-independent parameter to quantify the effect of lubrication on the tablet physical properties [1, 2]. Kushner [3] proposed empirical formulas, derived through experimental knowledge, to calculate the K-value.

Given the straightforward nature of the empirical formula [3], we adopted the K-value approach to scale-up a batch blending operation. We assume that K-value can be used to ensure both comparable lubrication and blend content uniformity across different scales. In this work, two case studies are presented from different stages of process development to explore the applicability of K-value to facilitate reliable scale-up of the blending operation for specific formulations.

The first application is at a later stage of process development when the expected variability in the material properties and the operating parameters relevant to blending are known. An extensive sensitivity analysis was used to identify the range of K-value expected over the range of variabilities in the blending parameters. Experiments were then designed at the extreme corners to confirm the K-value range.

The second application is from an early stage of process development, where an optimization routine was setup to design the blending process at the pilot and commercial scales based on the lab scale data to obtain maximum flexibility for process scale-up.

References:

[1] Kushner, J., & Schlack, H. (2014). Commercial scale validation of a process scale-up model for lubricant blending of pharmaceutical powders. *International Journal of Pharmaceutics*, *475*(1-2), 147-155.

[2] Nassar, J., Williams, B., Davies, C., Lief, K., & Elkes, R. (2021). Lubrication empirical model to predict tensile strength of directly compressed powder blends. *International Journal of Pharmaceutics*, *592*, 119980.

[3] Kushner IV, J. (2012). Incorporating Turbula mixers into a blending scale-up model for evaluating the effect of magnesium stearate on tablet tensile strength and bulk specific volume. *International Journal of Pharmaceutics*, *429*(1-2), 1-11.

184e Precipitant Surprise! Investigation and Mitigation of Media Precipitation during Scale-up and Technology Transfer

Benjamin Hawkins, Brian Thatcher, Caitlin Wasilko, Dixit Raiyani

Biopharmaceutical media powders are increasing in complexity to aid in increased cell densities and provide muti-cell platform utility. When hydrated, these complex powders produce high concentration solutions that are composed of dissolved and dispersed raw materials with difficult to predict solubility limits, complex chemistries, and challenging hydration procedures. Precipitation issues can arise or be discovered during technology transfer due to variations in hydration procedure and/or increased visibility at scale. Small-scale models for media hydration, coupled with chemical characterization techniques were used to identify the root cause of media precipitation. The isolated precipitates were analyzed by FTIR and EDX methods to understand their chemical characteristics. Experiments were performed to replicate the media precipitate formation, to different levels, by varying several parameters of the media hydration process and capturing their impact on turbidity, osmolality, component concentration via Cedex BioHT, Focused Beam Reflectance Measurement (FBRM), and Charged Aerosol Detection (CAD) signals throughout the hydration process. The small-scale studies and DOEs helped identify several hydration process parameters that impacted media precipitate formation. Implementation of the appropriate set of conditions for these parameters at scale resulted in a reduction in the amount of precipitation observed.

Viral Filtration Optimization for Challenging Non-mAb Exhibiting Rapid Flux Decay

Daniel Gendy, Peter Corbett Liu, Samantha San, Lauren Rockwell
Biologics Process R&D, Merck & Co., Inc., Rahway, NJ, USA

Abstract

The removal or inactivation of potential viral contaminants during the manufacture of therapeutic proteins is a critical component of the downstream purification process. Regulatory agencies typically recommend a minimum of two dedicated orthogonal viral clearance steps for biologically derived therapeutics. The diverse nature of viruses, such as variations in envelopment, necessitates the implementation of orthogonal viral clearance steps to decrease the likelihood of residual viral contaminants and increase product safety. Aside from a low pH viral inactivation, **viral filtration (VF) uses a porous membrane to achieve viral clearance by size-exclusion. This poster presents a case study on VF optimization of a challenging, highly potent non-mAb protein that exhibited significant fouling at platform viral filtration conditions.** The observed behavior necessitated substantial additional development beyond the standard VF platform fit, including a comprehensive **evaluation of buffer composition, pH, conductivity, excipient concentration, and filter chemistry impact on flux, loading, yield, and filtrate product quality**. Four different commercially available filters were screened at small-scale for loading prior to fouling. The Planova 20N filter was selected as the filter of choice due to the achievement of higher loading prior to fouling and increased yield. The addition and identity of excipients as well as buffer pH and conductivity were critical factors in mitigating flux decay, while the buffer matrix had minimal impact. **The optimized conditions resulted in a 50% improvement in VF filter loading compared to the starting platform settings and successfully mitigated adverse product quality impact during the VF step. These conditions enabled successful scale up of the VF step from bench to pilot and clinical scales.** By confirming a VF screening methodology for a non-platform hazardous product, this work can be leveraged to streamline VF development on future challenging atypical molecules.

Background

Table 1: Viral Filter Sizes and Compositions

Filter	Nominal Pore Size (nm)	Composition	Filter Sizes (m^2)	Operating Pressure Target (psi)
Viresolve Pro	18	**(a) Mixed Cellulose Ester**	0.0003	30
Virosart CPV	20	**(b) Polyether sulfone**	0.0005	25
Planova BioEx	20	**(c) Polyvinylidene Difluoride**	0.001	25
Planova 20N & S20N	20	**(d) Cuprammonium Regenerated Cellulose**	0.001	14 & 28

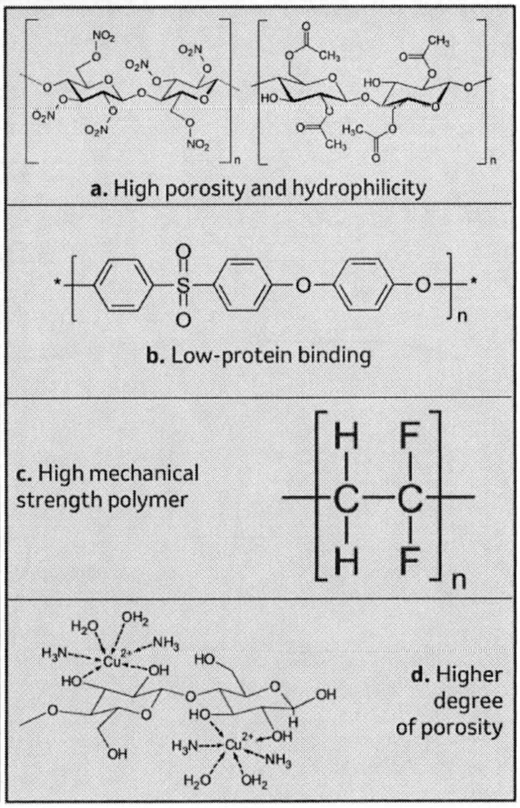

Figure 1: Viral Filter Schematic

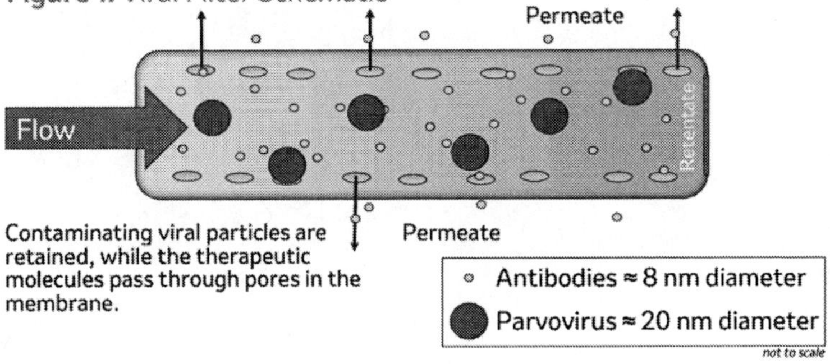

- Contaminating viral particles are retained, while the therapeutic molecules pass through pores in the membrane.

AIChE Annual Meeting, San Diego, CA, October 2024

Experimental Overview

Table 2: Study Setup

Title	Conditions Tested	Filters Used
Flush Buffer Studies	**1.** Glutamate Buffer, pH 5, 6 mS/cm	Planova 20N, Planova BioEx
	2. Histidine Buffer, pH 5, 6 mS/cm	Planova 20N, Planova BioEx, ViroSart CPV
Load Concentration pH 7 Study	**3.** 1x, 2x, 5x Load dilution, pH 7, 6 mS/cm	Planova 20N
L-Arginine & Salt Spike Load Conditioning Studies	**4.** 10 mM L-Arginine spike & Load dilution, pH 5, 6 mS/cm	Planova 20N, Viresolve Pro
	5. 50 mM & 100 mM L-Arginine spike, 6 mS/cm, Salt spike, 15 mS/cm, pH 5	Planova 20N
Low Pressure & pH/L-Arg Exploratory Studies	**6.** Reevaluation of No L-Arginine spike and 10 mM L-Arginine at lower average pressure	Planova 20N
	7. 150 mM L-Arginine spike, pH 4.5 & 8, 8.5 mS/cm	Planova 20N, Planova S20N

- Viral Filtration is typically performed after the final chromatography polishing step. In the case of these experiments the load material for the VF unit operation was product from cation exchange chromatography (CEXP).
- Concentration of the load material was typically around 5 g/L, unless dilution indicates otherwise
- Viral filters were operated within a biosafety cabinet in dead-end filtration mode using a pump adjusted for constant pressure feedback control with 0.5 L of load material prepared, targeting a loading of ≥ 400 L/ m^2
- The choice prefilter for all runs was Millipore SHR (0.5/0.1 μm)

Results

Figure 2: VF Screen: Studies 1 & 2 – Flush Buffer & Filter Screen

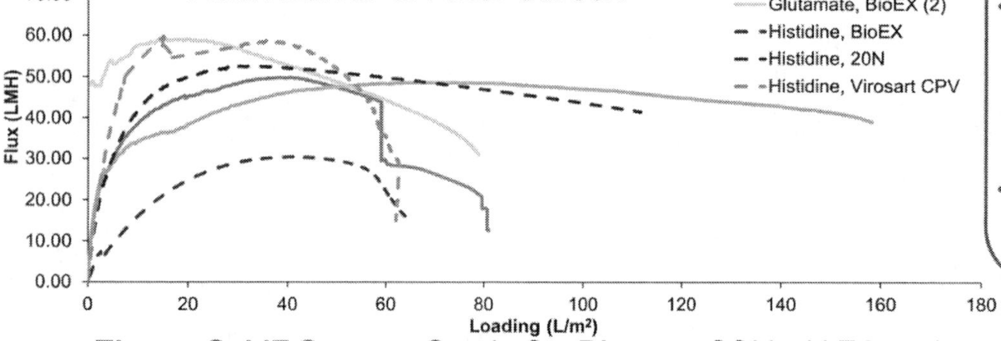

- Flush Buffers at pH 5, 6 mS/cm.
- Found no direct impact on the loading or flux with the choice of flush buffer matrix use → Sodium Acetate, pH 5 selected for alignment benefit with previous unit operation.
- Differences in filter performances → Planova 20N selected (loads higher, slower decay)

Figure 3: VF Screen: Study 3 – Planova 20N pH 7 Load Dilutions

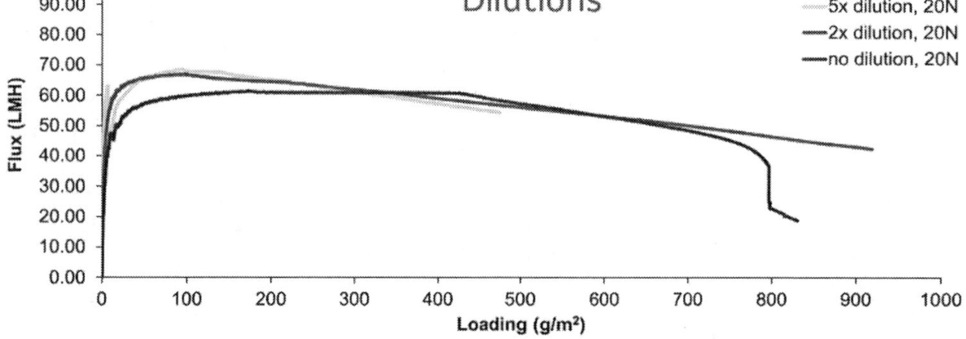

- Load and flush buffers adjusted to pH 7, 6 mS/cm
- No clear benefit of diluted loads (impractical due to long run times)
- pH 7 load reached higher initial flux yet rapid flux decay starting at ~75 L/ m^2 (800 g/ m^2) → pH 5 condition maintained

Figure 4: VF Screen: Studies 4 & 5: L-Arg & High Conductivity Spike

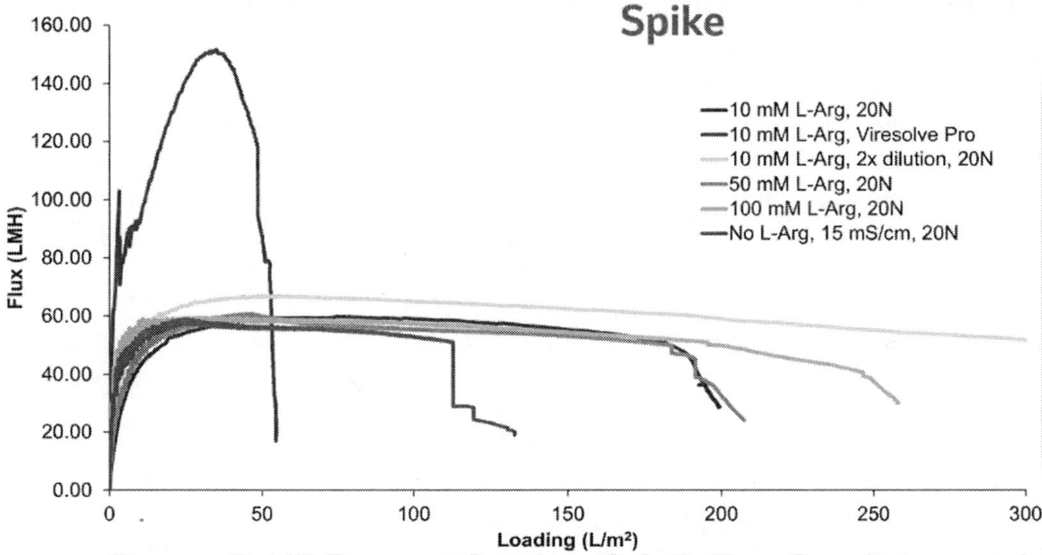

- Significant increase in loading with the presence of L-Arg as compared to the no L-Arg conditions.
- Improvement in loading was robust across a range of L-Arg additions (10mM to 100mM).
- Higher conductivity had no benefit and resulted in lower loading than the 6 mS/cm loads previously used.

Figure 5: VF Screen: Studies 6 & 7: Top Conditions at Low Pressure + Exploratory Runs

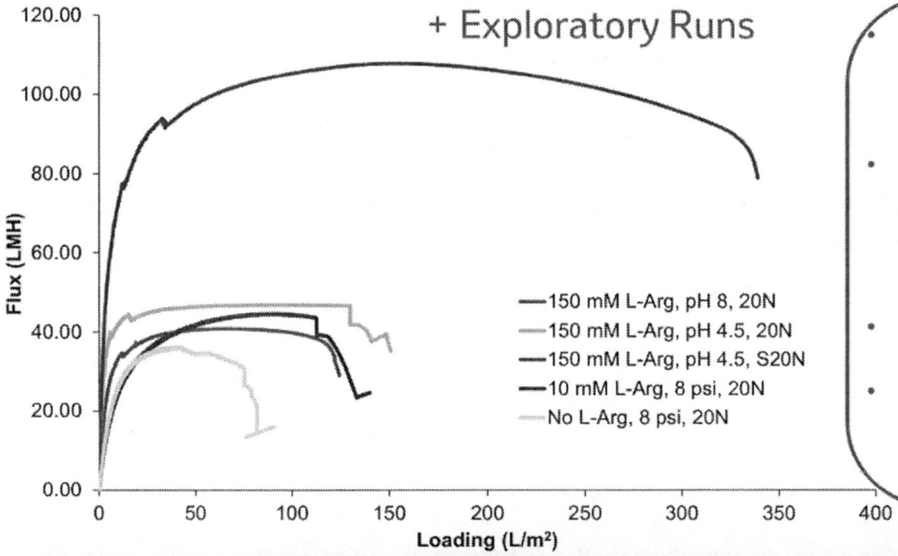

- Low Pressure unfavorable due to slower filtration rates (lower flux → increased aggregation) and potential for virus escape from pores.
- Higher concentrations of L-Arg spike slightly improved loading. L-Arg solution is highly basic, 10 mM minimizes impact on VFL and cost at scale while maximizing throughput.
- Again, no significant difference between high and low pH
- S20N more than doubles the throughput and flux observed from the 20N (supply was unavailable with scale up timelines)

Conclusions

- Unadjusted viral filtration loads resulted in flux decay beginning at an average of approximately 100 L/m^2 or 550 g/m^2 with platform conditions. Proceeding with conditions identified for scale up (10 mM L-Arg, pH 5, 6 mS/cm, Planova 20N) resulted in a throughput of 175-200 L/m^2 or 900-1000 g/m^2, an improvement of ≥ 50 %.
- Addition of L-Arginine to the VF load resulted in increased product throughput. Supports claims of arginine as a versatile solvent for protein-protein aggregation suppression and favorable interactions with protein amino acid side chains **(3).**
- Selected conditions demonstrate no significant change in product purity across viral filtration unit operation (~98.5% purity) and resulted in successful scale up to pilot and clinical batches (~93% yield).
- Performance can be improved with next generation viral filters (S20N).

Table 1d. Wurm, Florian & Rietzler, Barbara & Pham, Tung & Bechtold, Thomas. (2020). Multivalent Ions as Reactive Crosslinkers for Biopolymers 25. 1840. 10.3390/molecules25081840.
Table 1 pore sizes. Fallahianbijan F, Giglia S, Carbrello C, Bell D, Zydney AL. Impact of protein fouling on nanoparticle capture within the Viresolve® Pro and Viresolve® NFP virus removal membranes. *Biotechnology and Bioengineering*. 2019; 116: 2285-2291. https://doi.org/10.1002/bit.27017
3. Arakawa T, Ejima D, Tsumoto K, Obeyama N, Tanaka Y, Kita Y, Timasheff SN. Suppression of protein interactions by arginine: a proposed mechanism of the arginine effects. Biophys Chem. 2007 Apr;127(1-2):1-8. doi: 10.1016/j.bpc.2006.12.007. Epub 2006 Dec 21. PMID: 17257734.

184h Particle Jamming upon Administration of Long-Acting Injectable Crystalline Drug Suspensions

Ziwan Xu, Kavin Kowsari, Kartik Bomb, Latevi Lawson, Faustin Arevalo, Ashley Johnson

Long-acting injectable (LAI) crystalline drug suspensions, administered via intramuscular (IM) or subcutaneous (SC) injection, allow sustained drug release over extended periods (weeks to months) following a single administration, greatly reducing dosing frequency and improving patient adherence. The development of crystalline suspensions requires consideration of multiple critical quality attributes, including physical and chemical stability of the suspension, homogeneous dispersion of particles for content uniformity and dose accuracy, and the ability to easily withdraw the suspension into a syringe and administer the formulation to tissue.

However, significant gaps still exist in establishing tools to appropriately characterize long-acting suspensions, particularly with respect to particle aggregation, sedimentation, and injectability to define an appropriate formulation design space. Here we used monodisperse, spherical polymer microparticles of different sizes and prepared different suspension formulations to study the influence of particle size distribution, particle density, drug loading, suspension viscosity and injection conditions on sedimentation and injectability. Specifically, sedimentation was monitored using *in-situ* probes, such as EasyViewer and Blaze, which tracked real-time particle size, counts and concentration. Subsequently, injectability was assessed by employing syringe pumps for injections into air or porcine tissue. Additionally, we used fluorescence microscope to capture the particle flow within the syringe body and needle, visualizing the process of particle jamming and analyzing the clogged state. Through parameter sensitivity analysis, we identified the appropriate formulation design space to mitigate particle jamming and ensure injectability of long-acting injectable suspensions.

184i Comparison Among Lnp, PNP and Exosomes for Gene Delivery to Treat Triple-Negative Breast Cancer

Tanvi Varadkar, Kai Chen, Jiashuai Zhang, Zhuoxin Zhou, Lufang Zhou, X. Margaret Liu

The gene therapy for cancer treatment needs efficient, stable and safe delivery vehicles. The goal of this study was to compare three gene delivery vehicles, i.e. cationic lipid nanoparticles (LNPs) using traditional DOTAP and DOPE, RALA peptide nanoparticles (PNPs), and exosome nanoparticles (Exo). Specifically, we synthesized or produced these three vehicles and characterized them with NanoSight, Zeta and TEM. The gene packing rate of our mitochondria-targeting mLumiOpto genes was compared using RT-PCR. The cancer specificity was tested by surface tagging our monoclonal antibody that was developed to target the surface receptor overexpressed in triple-negative breast cancers (TNBCs). Then, we evaluated the transfection/transduction efficiency and functional gene expression using multiple TNBC cell lines. Finally, the *in vivo* cancer targeting and gene delivery were evaluated and compared. Taken together, this study provided a full comparison of three nanoparticles for cancer targeting and gene therapy.

184j Improving Numerical Methods for Bioreactor Process Simulation and Parameter Estimation

Taehun Kim, Miguel Angel Valderrama-Gomez, William Heymann III, Juliane Glaser, Matthew Demers, Sidra Xu, Elcin Icten Gencer, Fabrice Schlegel, Pablo Rolandi

This abstract discusses methods to speed up numerical integration of Ordinary Differential Equations (ODEs) and optimization for Flux Balance Analysis (FBA) in bioreactor models, aiming to improve the numerical performance of search-based parameter estimation algorithms.

In silico mammalian cell culture bioreactor modeling platforms hold great promise for transforming bioprocess development. Simulation and optimization of biomanufacturing processes can reduce experimentation and enhance process efficiency, as well as product quality. Building a predictive dynamic model requires estimating parameters based on bioreactor experimental data. Parameter estimation can represent a significant computational challenge because it requires solving a high dimensional non-convex optimization problem to an acceptable local or global optimality. Performing a successful parameter estimation requires practitioners to carefully formulate the optimization problem and to select efficient numerical methods for integration and optimization. This poses a significant risk of identifying sub-optimal solutions if an inappropriate parameter estimation workflow is employed.

In this work, we present two approaches to significantly speeding up the numerical integration of ODEs, as well as the numerical optimization for solving FBA, involved in simulating in-house developed bioreactor models for describing cellular metabolisms. First, an idea of better scaling the problem has been explored. To do so, the underlying equations for a model with the cellular metabolism described with simple kinetic equations (e.g., see [1]) were nondimensionalized. Thanks to the normalization of the model variables and parameters, numerical solvers can take advantage of better scaling of the problem, thus resulting in a much better performance. To demonstrate this concept, we compare the numerical performance of the dimensional and dimensionless simulations of the simple kinetic model. Second, alternate formulations for FBA in a hybrid model (e.g., see [2]), as well as the corresponding convex optimization solvers from the CVXOPT library (see [3]), have been explored. Without loss of generality, FBA can be formulated either as a Quadratic Program (QP) or a Second Order Cone Program (SOCP). To test each solver's numerical performance, solution time and accuracy, and stability of the obtained solutions were compared within the hybrid model's context.

Since our model parameter estimation algorithm is based on a genetic algorithm, which couples an ODE solver with a search-based numerical optimization algorithm, speeding up the numerical integration of ODEs, as well as the numerical optimization for solving FBA, holds a promise in achieving a significant reduction of the model parameter estimation time.

[1] Kontoravdi, C, Pistikopoulos, E. N., and Mantalaris, A. Systematic development of predictive mathematical models for animal cell cultures. Computers & Chemical Engineering. 2010;34(8):1192–1198. doi:https://doi.org/10.1016/j.compchemeng.2010.03.012.

[2] Nolan, R. P. and Lee, K. Dynamic model of CHO cell metabolism. Metabolic Engineering. 2011;13(1):108–124. doi:https://doi.org/10.1016/j.ymben.2010.09.003.

[3] Andersen, M., Dahl, J., and Vandenberghe, L. CVXOPT: Python software for convex optimization, version 1.2. http://cvxopt.org/, February 2019.

184k Gaussian Process Regression Modeling to Predict α/β-Peptide Selectivity for Antifungal Drug Development

Joshua Richardson, Douglas Chang, David M. Lynn, Sean P. Palecek, Reid Van Lehn

Since the introduction of insulin to treat type 1 diabetes 100 years ago, large strides have been made in the field of peptide therapeutics to treat a range of other diseases such as cancer, multiple sclerosis, and HIV.[1] A pressing issue in medicinal research today is leveraging the knowledge of current peptide synthesis techniques to design antifungal drugs. Given the recent development of drug-resistant strains to current treatments on the market, there is large interest in developing antimicrobial peptides (AMPs) that work via membrane disruption and are therefore less likely to lead to resistance.[2] However, naturally sourced α-helical AMPs have low stability in vivo due to their susceptibility to proteolytic cleavage, and they exhibit low selectivity for fungal versus human cells.[3] Consequently, new approaches to design synthetic peptides are needed to tackle these challenges.

One promising class of synthetic peptides is α/β-peptides, where α denotes traditional amino acids while β denotes modified amino acids with an additional backbone carbon. An advantage of these materials is the large array of α/β combinations that can be tested, each with differing stability profiles, distributions of side chains, etc. The goal in designing α/β-peptides for antifungal applications is to strike a balance in maximizing the activity against fungal pathogens – quantified as decreasing the minimum inhibitory concentration (MIC) of peptide required to prevent pathogenic cell growth – while minimizing hemolysis of human red blood cells – quantified as increasing the peptide concentration required to lyse 10% of red blood cells (HC_{10}).[4] Maximizing the ratio of HC_{10} to MIC through a metric known as the Selectivity Index (SI) is desired during the rational development of new antifungal peptides.[3,4] Recent experimental studies have shown that only a handful of measured quantities such as helicity and liquid chromatography retention time can be utilized to accurately predict fungal activity and hemolysis metrics.[3,5] However, these experiments require α/β-peptide synthesis and consequently have a large time and financial cost, limiting their utility in large-scale model development.

To address this challenge, we hypothesized that computationally derived descriptors could be utilized to more quickly and accurately predict HC_{10} and MIC measurements prior to α/β-peptide synthesis. Using a dataset of 150 published α/β-peptide sequences as a starting point, we built a Gaussian Process Regression (GPR) workflow to predict HC_{10} and MIC values by calculating 2D molecular descriptors with the RDKit cheminformatics toolkit and simplified SMILES string representations of α/β-peptide sequence. We developed and implemented an iterative model training approach to predict new peptide experimental labels using a design space motivated by the most selective (highest SI) sequences in the initial training set. Peptides were selected for synthesis based upon GPR predictions, with measured HC_{10} and MIC values then used to update GPR model parameters to guide the next round of predictions. The advantage of this workflow is that GPR provides estimations of uncertainty for prospective sequences, allowing us to probe both promising new sequences with low uncertainty and high predicted SI while also testing sequences with high uncertainty to expand the design space (e.g., new amino acids) for future rounds. We show that this approach identifies new selective α/β-peptides and sequence features important to HC_{10} and MIC predictions. These results demonstrate the potential of this approach to screen for new highly selective sequences for peptide drug development.

(1) Muttenthaler, M.; King, G. F.; Adams, D. J.; Alewood, P. F. Trends in peptide drug discovery.

Nature Reviews Drug Discovery **2021**, *20* (4), 309-325. DOI: 10.1038/s41573-020-00135-8.

(2) Benfield, A. H.; Henriques, S. T. Mode-of-Action of Antimicrobial Peptides: Membrane Disruption vs. Intracellular Mechanisms. *Frontiers in Medical Technology* **2020**, *2*, Mini Review.

(3) Lee, M.-R.; Raman, N.; Gellman, S. H.; Lynn, D. M.; Palecek, S. P. Incorporation of β-Amino Acids Enhances the Antifungal Activity and Selectivity of the Helical Antimicrobial Peptide Aurein 1.2. *ACS Chemical Biology* **2017**, *12* (12), 2975-2980. DOI: 10.1021/acschembio.7b00843.

(4) Murugan, R. N.; Jacob, B.; Ahn, M.; Hwang, E.; Sohn, H.; Park, H.-N.; Lee, E.; Seo, J.-H.; Cheong, C.; Nam, K.-Y.; et al. De Novo Design and Synthesis of Ultra-Short Peptidomimetic Antibiotics Having Dual Antimicrobial and Anti-Inflammatory Activities. *PLOS ONE* **2013**, *8* (11), e80025. DOI: 10.1371/journal.pone.0080025.

(5) Chang, D. H.; Lee, M.-R.; Wang, N.; Lynn, D. M.; Palecek, S. P. Establishing Quantifiable Guidelines for Antimicrobial α/β-Peptide Design: A Partial Least-Squares Approach to Improve Antimicrobial Activity and Reduce Mammalian Cell Toxicity. *ACS Infectious Diseases* **2023**, *9* (12), 2632-2651. DOI: 10.1021/acsinfecdis.3c00468.

184l High-Throughput Experimentation Reveals Scope and Limitations of Selective Phosphine Reductants and Enables One-Pot Mab Reduction/Conjugation

Cuixian (Trisha) Yang, Marion Emmert, Eugene Kwan, Rebecca A. Chmielowski, Bruce Kilgore, Zachary Vanaernum, Cecilia Bottecchia, Rodell Barrientos, Monica Haley, Kelly Raymond, Michael Rauscher, Zachary D. Dunn, Jay Desai

This work describes the detailed evaluation of more than 40 phosphine reductants via automated and non-automated high-throughput experimentation approaches with the goal to identify selective reductants for cleaving the monoclonal antibody (mAb) disulfide bonds of capped, engineered cysteines. As a point of reference, this study included phosphines that have previously been documented in the literature [4-diphenylphosphino benzoic acid (DPPBA), tris(3-sulfophenyl)phosphine (TSPP), and 3-(diphenylphosphino)benzenesulfonate (diPPBS)] but showed relatively poor selectivity. The high-throughput experiments performed in this study revealed several phosphines with potential for selective reduction, which have not been previously studied for this type of transformation. These initial hits are further evaluated with regard to phosphine/mAb ratio, solubility in aqueous media, and air oxidation behavior. The best phosphine identified through these studies (1-(4-(diphenylphosphanyl)benzyl)-1-methylpyrrolidin-1-ium bromide) was then employed in a high-throughput study that established efficient one-pot reduction/conjugation reaction conditions. Overall, the studies detailed herein demonstrate how high-throughput experimental design enables rapid and resource-sparing insights into mAb reduction and conjugation reactivity with phosphine-based reductants.

184n Thermodynamics Analysis with Molecular Informatics for Understanding Pharmaceutical Cocrystallization Promoted in Fatty Acid Media

Yuna Tatsumi, Yasuhiko Orita, Yusuke Shimoyama

Introduction

Pharmaceutical cocrystal is one of the crystal engineering techniques to improve the physical-chemical properties of active pharmaceutical ingredients (APIs) such as solubility, bioavailability and stability. [1] Cocrystal is a crystal composed of API and other molecules called as coformer. The crystal structure of cocrystal maintains the non-covalent bond between API and coformer, which allows the cocrystallization technique to be applied to API and coformer with no dissociable group. To achieve the development of pharmaceutical cocrystal, it is necessary to design both suitable combination of API and coformer and cocrystallization media.

Liquid-assist grinding has been known for the representative cocrystallization method. In liquid-assist grinding, the API and coformer are stirred with small amounts of organic solvent. The cocrystal formation depends on the kinds of solvent added. [2] The toxic organic solvents are sometimes used as auxiliary to promote cocrystallization. However, the use of toxic organic solvents is a serious issue that they may remain in the products.

To overcome the issue of liquid-assist grinding, our group has developed the cocrystallization method to use liquefied fatty acid as auxiliary species, instead of organic solvents [3]. Fatty acid is a safe material and generally included in oils. Linoleic acid (LA), which was one of the general fatty acids, was applied to the theophylline (TPL) cocrystallization [3]. Our proposal method achieved TPL cocrystallization by the addition of LA successfully, but our method remains an issue that the effect of fatty acid on cocrystallization is still unknown. Therefore, this research aims to analyze the effect of fatty acid on cocrystallization from both experiment and modelling.

Methods

[Experiment] TPL (1 mmol) as API and nicotinamide (NA, 1 mmol) as coformer were added to the vessel. Fatty acid and hydrocarbon are used as liquid auxiliaries. After adding LA (0.4 g) as fatty acid or 1-octadecene (1-OD, 0.4 g) as hydrocarbon used reference, the mixture was stirred during 120 min under 22 ^{o}C. The crystal structure was evaluated by Powder-Xray diffraction (PXRD). The ratio of cocrystal formation was also evaluated as the ratio of the characteristic peak intensity of cocrystal to the sum of characteristic peak intensities of API, coformer and cocrystal.

[Modelling] To investigate the effect of cocrystallization promotion given by liquid auxiliaries, the cocrystallization under liquid auxiliaries is composed of the four phases: (A) exist as raw materials (B) active API and coformer by auxiliaries, (C) make the pair and (D) crystallize them. We focused on the phase B in this work because the phase B seems to be the phase that receives the influence of liquid auxiliaries. The energy of each phase A and B was calculated by thermodynamics and molecular information [4].

Results & Discussion

[Experiment] The crystal structures were evaluated for products that were fabricated by stirring TPL and NA with LA and 1-OD by PXRD measurements. The peak attributed to cocrystal [5] of the product stirred with LA was bigger than that of the product stirred with 1-OD where the peaks attributed to raw materials of the product stirred with LA were smaller than those of the product stirred with 1-OD. The cocrystallization ratio of the product using LA was much larger than that of the product using 1-OD. These results mean that LA plays a role to enhance cocrystal composed of TPL and NA more than 1-OD.

[Modelling] The Gibbs free energy difference between the phase A to B (ΔG_1) was calculated by thermodynamic model combined with molecular information. ΔG_1 was smaller with LA as auxiliaries than that with 1-OD. The auxiliary whoseΔG_1 was small value means that the fatty acid assist the formation of API and coformer pair. The result of ΔG_1 calculation means that LA is more suitable auxiliary to active raw materials than 1-OD. It was also confirmed that the addition of LA promoted the cocrystallization than that of 1-OD from the experimental results. Therefore, the developed model is expected to estimate the effect of liquid auxiliary on the promotion of cocrystallization.

Conclusion

We investigated the effect of fatty acids on cocrystallization from both experiment and modelling. From experiments, the addition of fatty acid promoted TPL cocrystallization in the comparison of hydrocarbon. Additionally, from modelling based on thermodynamic model combined with molecular information, fatty acid may activate raw materials more than hydrocarbon. These results mean that fatty acid is more suitable auxiliary to form API and coformer pair promote their cocrystallization than hydrocarbon. Therefore, the effect of auxiliary can be explained by the model based on thermodynamic model combined with molecular information, which is expected that the suitable auxiliary is estimated without any experiments.

References

[1] Sithmi, N. M.; Aruna M.; Ranjit T.; Nadeesh M. A. Org. Process Res. Dev. 27 (2023) 409−422.

[2] David R. W.; Tanise S.; Peddy V.; Michael J. Z. Cryst. Growth Des. 9, (2009) 1106−1123.

[3] Yuna T.; Yasuhiko O.; Yusuke S.; Cryst. Growth Des. 22, (2022) 5176–5181.

[4] Shiang-Tai L.; Stanley I. S.; Ind. Eng. Chem. Res. 41 (2002) 899-913.

[5] Jie, L.; Sohrab, R. Preparation and Characterization of Theophylline-Nicotinamide Cocrystal. Org. Process Res. Dev. 2009, 13, 1269−1275.

184o Risk-Based Quality Control for the Safe and Efficient Biotechnological Batch Production of Lactic Acid

Austin Braniff, Dylan Miller, Beatriz Dantas, Sahithi Srijana Akundi, Yuanxing Liu, Shayan Niknezhad, Faisal Khan, Efstratios Pistikopoulos, Yuhe Tian

Process control plays a critical role in pharmaceutical fed-batch and batch bioreactors to ensure safe and high-quality products [1-2]. More than 90% of the bioprocessing industry still relies on conventional error-reactive single-input single-output (SISO) control techniques, which may identify sub-optimal control actions given the complex system dynamics under potential disturbances in pH, temperature, nutrient intake, etc. [3-4] Advanced model-based control is thus essential to enhance the operations in these bioprocesses, such as model predictive control (MPC) which leverages repetitive online dynamic optimization to determine the optimal control actions [5-6]. However, key research questions remain on: (i) How to simultaneously account for process control and batch-end quality control? (ii) How to ensure the resulting optimal process operating pathway is safe? and (iii) How to address online computational efficiency?

In this work, we present a systematic approach to integrate risk-based control and end-point quality control for pharmaceutical processes using the fermentation production of lactic acid as a case study. We first discuss the data-driven surrogate modeling for the control of this process based on a verified high-fidelity model [7]. An explicit MPC problem is then formulated using the surrogate model to maintain the process risk within an acceptable level [8]. Herein, risk is defined based on the historical operating data of the process. For example, the temperatures or pH levels at which the substrate may begin to degrade, or the high pressures at which the reactor vessel may be damaged. These safety-critical parameters are holistically evaluated in real time quantified via a dynamic risk index [9]. A multi-level optimization problem is then formulated to incorporate both closed-loop risk control and end-point quality control which has been demonstrated on a chemical batch production process [10]. This optimization problem is recast as a multi-parametric problem and solved offline to obtain explicit solutions to the quality control problem. These solutions are tabulated as functions of measurable process parameters so that the optimal bioreactor control input can be quickly evaluated in real-time to ensure batch quality is met in a safe fashion. Moreover, online optimization can be replaced with function evaluation to significantly reduce computational load during actual operations. This approach is showcased to identify the safely optimal operating trajectory of the lactic acid production process while satisfying the end-batch quality target.

References

1. [1] Jagschies, G., Lindskog, E., Łacki, K., & Galliher, P. (2018). Biopharmaceutical Processing: Development, Design, and Implementation of Manufacturing Processes. In Biopharmaceutical Processing: Development, Design, and Implementation of Manufacturing Processes (p. 1275).
2. [2] Melgarejo-Torres, R., Pérez-Vega, S. B., Rivera-Arredondo, V. M., & Che-Galicia, G. (2019). Chapter Six – Multiphase bioreactors in the pharmaceutical industry. In S. Huerta-Ochoa, C. O. Castillo-Araiza, & G. Quijano (Eds.), Advances in Chemical Engineering (Vol. 54, pp. 195–237). Academic Press.
3. [3] Mitra, S., & Murthy, G. S. (2022). Bioreactor control systems in the biopharmaceutical industry: A critical perspective. Systems Microbiology and Biomanufacturing, 2(1), 91–112.
4. [4] Murthy, G. S., Johnston, D. B., Rausch, K. D., Tumbleson, M. E., & Singh, V. (2012). Design and Evaluation of an Optimal Controller for Simultaneous Saccharification and Fermentation Process. Applied Biochemistry and Biotechnology, 166(1), 87–111.

5. [5] Luo, Y., Kurian, V., & Ogunnaike, B. A. (2021). Bioprocess systems analysis, modeling, estimation, and control. Current Opinion in Chemical Engineering, 33, 100705.
6. [6] Benattia, S. E., Tebbani, S., & Dumur, D. (2020). Linearized min-max robust model predictive control: Application to the control of a bioprocess. International Journal of Robust and Nonlinear Control, 30(1), 100–120.
7. [7] Gordeeva, Yu. L., Rudakovskaya, E. G., Gordeeva, E. L., & Borodkin, A. G. (2017). Mathematical modeling of biotechnological process of lactic acid production by batch fermentation: A review. Theoretical Foundations of Chemical Engineering, 51(3), 282–298.
8. [8] Ali, M., Cai, X., Khan, F. I., Pistikopoulos, E. N., & Tian, Y. (2023). Dynamic risk-based process design and operational optimization via multi-parametric programming. Digital Chemical Engineering, 7, 100096
9. [9] Bao, H., Khan, F., Iqbal, T., Chang, Y. (2011). Risk-based fault diagnosis and safety management for process systems. Process Saf. Prog., 30(1), 6-17.
10. [10] Braniff, Y. Tian, "Dynamic Risk-based Model Predictive Quality Control with Online Model Updating," American Control Conference, July 8, 2024.

184p Impact of Droplet Evaporation on Polymorphism of Suberic Acid

Sheriff Anofi, Ryan Snyder

In the diverse landscape of the pharmaceutical sector, which presents its own mix of challenges and opportunities, the significance of polymorphism is highlighted as the industry continues to expand and active ingredients become more complex. Polymorphism in a drug compound signifies its capacity to adopt various crystal forms. Variations in polymorphic forms of drug compounds can significantly impact bioavailability, solubility, stability, and efficacy, posing substantial challenges in drug formulation and regulatory compliance. To overcome these challenges and enhance drug development and formulation, implementing a polymorph screening process is essential. This process involves identifying and characterizing the different solid-state forms that a drug molecule can adopt, including various crystal structures, hydrates, and amorphous forms. Despite the high and consistent loss rate of drug candidates, conducting thorough screening is costly and time-consuming. However, it is crucial to select the appropriate polymorph to address stability and physicochemical properties. Carboxyl group-containing organic compounds are extensively used across a wide range of applications within the pharmaceutical industry. This work aims to investigate the polymorphism of suberic acid ($C_8H1_4O_4$), a dicarboxylic acid utilized in drug synthesis and as a pharmaceutical co-former, with a specific focus on the influence of crystallization conditions, including solvent choice, concentration all using a droplet evaporation technique. While suberic acid has been studied extensively its polymorphism has been only a more recent focus with only two polymorphs of suberic acid identified. In this work, we present a potential third polymorph that is formed through droplet evaporation.

A constant output aerosol generating atomizer is used to create a microparticles of suberic acid from solution via droplet evaporation at room temperature. Polydisperse droplets containing suberic acid particles are subjected to drying in a series of dryer columns to evaporate the solvent. In this study, various solvents including water, ethanol, and isopropyl alcohol were used at different droplet concentrations. The impact of variations in crystallization conditions on suberic acid polymorph formation was investigated to determine the conditions for successful polymorph isolation. Suberic acid particles undergo analysis utilizing a combination of techniques. X-ray Diffraction (XRD) examines the internal crystalline structure via diffraction patterns, while Scanning Electron Microscopy (SEM) investigates sample morphology, including size, shape, and surface characteristics. General Scanning Mobility Particle Size (SMPS) determines particle size distribution, and Differential Scanning Calorimetry (DSC) analyzes polymorphic transformations and thermal transitions.

184q Development of a Lab-Scale Fluid Bed Granulation Process within a Semi-Continuous Manufacturing Workflow

Sydney Butikofer, Marly Jo Santos Colón, Joshua Hanson, Jonathan Brett Wade, Zhanjie Liu, Shashwat Gupta

1) Purpose

Fluid bed wet granulation (FBG) is a common pharmaceutical unit operation where a powder blend is fluidized, sprayed with a binder solution, and dried within a single process vessel. The purpose of including a granulation step in a drug product manufacturing process is to improve physical properties of an active pharmaceutical ingredient (API) by agglomerating polydisperse particles into a uniform size distribution. While FBG is typically operated as a batch process, semi-continuous manufacturing processes have recently emerged which offer a hybrid option offering the well-established advantages of both batch and continuous processing techniques [1]. Recently, a semi-continuous fluid bed granulation system has been introduced commercially (Xelum®, Syntegon, Germany), with broadly three iterations for consideration (Figure 1): lab-scale (SolidLab 1 -SL1), commercial scale 1-chamber (Xelum R&D), and a commercial scale 5-chamber. The SL1 is envisioned as a material sparing lab-scale benchtop fluid bed granulator for formulation prototyping and process parameter screening. The 1-chamber SolidLab 2 (SL2) is envisioned for at-scale process development, clinical trial and initial commercial production. The 5-chamber is positioned as a large-scale commercial manufacturing platform.

In pharmaceutical development, implementation of a lab-scale FBG for early formulation development poses several advantages. First, it allows for the reduction of total API used during development which is often a limiting resource. Second, lab-scale FBG may be used to evaluate prototype formulations for CQA screening. Finally, process and formulation variables may be characterized via appropriate experimental designs, while reducing time and material investment.

The objective of this work is to develop a comprehensive process understanding of a lab-scale FBG. First, a representative batch size was established utilizing prior experience, vendor recommendations, and preliminary ranging studies. The lab-scale FBG was further characterized at the established batch size by a statistical design of experiments (DOE) based approach to investigate the impacts of formulation attributes, operational parameters, and design variables on relevant desired quality attributes of granules and tablets. In addition, insights gained from the DOE were used to evaluate the feasibility of a dimensionless scaling criterion, the evaporative energy to drying capacity (EE/DC) ratio [2], to maintain granule characteristics across granulation scales (lab to commercial). Finally, a multiway partial least squares model (MPLS) was constructed to characterize the dynamic impact of process variables on granule characteristics.

2) Methods

This study delves into four APIs (micronized Acetaminophen or APAP, LY-1, LY-2, and LY-3) distinguished by their distinct particle sizes, bulk densities, and flowabilities. These APIs were selected for this analysis to demonstrate process robustness across a range of typical material-intrinsic failure modes observed at high drug load formulations for a direct compression process. Based on previous prototyping evaluations, formulations containing a high drug load of LY-1 and compressed via direct compression (DC) produced tablets with low tensile strengths. In addition, formulations containing high drug loads of LY-2 demonstrated poor flow and the corresponding tablets showed edge chipping and picking/sticking. Finally, LY-3 demonstrated prototyping challenges with poor flow and corresponding tablet weight variation when processing via DC or roller compaction (RC). Each formulation contained a drug load greater than or equal to 40%, a filler (MCC), a binder (HPC), and a disintegrant (CCS). Small amounts of extragranular excipients (Syloid, SSF, or MgSt) were added as needed (<6.5%).

Granulation operating parameters were varied in an initial ranging study to observe the failure modes of the FBG process. Once a successful operating space was defined, granules were tested for particle size distribution (PSD) to identify a target set of conditions that produced desirable granules representative of the commercial process. Granules were tested for PSD, bulk and tapped density, final moisture content, and flow properties (via shear cell testing).

The selected final blends were compressed into tablets using an HB50 compaction simulator (Huxley Bertrum Engineering Ltd., Cambridge, UK) simulating a Korsch XL200 rotary tablet press at 90 rpm. TSM B round flat face punches with a 6mm diameter were employed. For each selected final blend, three compression pressures were applied, resulting in tablets with different solid fractions. Tablets were tested for weight, thickness, and hardness measurements, as well as disintegration times.

3) Results & Discussion

A preliminary set of experiments was conducted to establish an initial operating space for further development. The goal of this work was to identify a value for EE/DC that produced a representative granulation at a 400 g batch size, such that the resulting granules possess similar characteristics to those from a typical commercial process. Subsequently, EE/DC was calculated for the successful runs and used for scaling down batch size to 200 g and 100 g. For a particular set of process conditions, this ratio indicates the amount of moisture in the granule bed during the granulation phase of an FBG process.

Granulations with 100 g batch sizes suffered from poor fluidization and an overall lack of reproducibility in granule size distribution. Thus, 200 g was selected as a minimum batch size for further work. Select runs at the 400 g and 200 g scale were evaluated for flowability and compactibility. Tablets from these granulations demonstrated good tensile strength (>2.0 MPa) and acceptable disintegration times (<5 minutes).

Next, three APIs (LY-1, LY-2, and LY-3) were granulated in place of micronized APAP in the formulation at the identified target conditions for the 200 g batch size. In all cases, the granules were blended with a lubricant (sodium stearyl fumarate) and compressed into tablets to evaluate the critical quality attributes. LY-1 tablets had an average tensile strength of 2.4 MPa at the nominal solid fraction (0.85). For LY-2, optical microscopy was performed on the compressed tablets to confirm that no edge chipping was observed and picking/sticking was significantly reduced. Upon comparison with previously made tablets, it was determined that the SL1 tablets resulted in better tablet compaction at high drug loads. Lastly, for LY-3, PSD results demonstrate acceptable particle size enlargement.

As a third objective, the SL1 process ranges for a 200 g batch size were defined to understand the quantitative effect of process (binder spray rate, inlet air temperature, spray air pressure) and design variables (nozzle air cap size) on granule attributes via DOE-based experimentation. For this study, extreme conditions were selected to simulate a high- and low-moisture accumulation/removal rate scenarios, which correspond to over-granulation and under-granulation conditions, respectively. Using the process data and granule responses captured from the 17-run DOE, a MPLS model was built using 4 latent variables and explains 90.8% of the variance in the granule size descriptors.

Among all variables in the process data studied, the spray air pressure was found to have the highest correlation with the granule attributes.

4) Conclusion

In this study, a lab-scale fluid bed granulation process was evaluated at multiple process conditions, batch sizes and formulations. A dimensionless scaling criterion, EE/DC, was used to maintain granule characteristics across batch size. Three high drug load formulations containing different APIs (LY-1, LY-2, and LY-3) showed improvements in their respective failure modes when granulated in the SL1 prior to compression. Furthermore, a 17-run DOE was executed to characterize the impact of process and design variables on the granule attributes. Among the variables studied, the spray air pressure was found to have the most significant effect on the granule size descriptors.

Figure 1. Process development workflow for semi-continuous fluid bed granulation

5) References

[1] Muzzio, F., & Oka, S. (Eds.). (2022). How to Design and Implement Powder-to-Tablet Continuous Manufacturing Systems. Academic Press.

[2] E. Gavi and A. Dischinger, 'Scale-up of Fluid Bed Granulation Using a Scale-Independent Parameter and a Process Model,' AAPS PharmSciTech, vol. 22, no. 4, p. 148, May 5 2021, doi: 10.1208/s12249-021-02013-x.

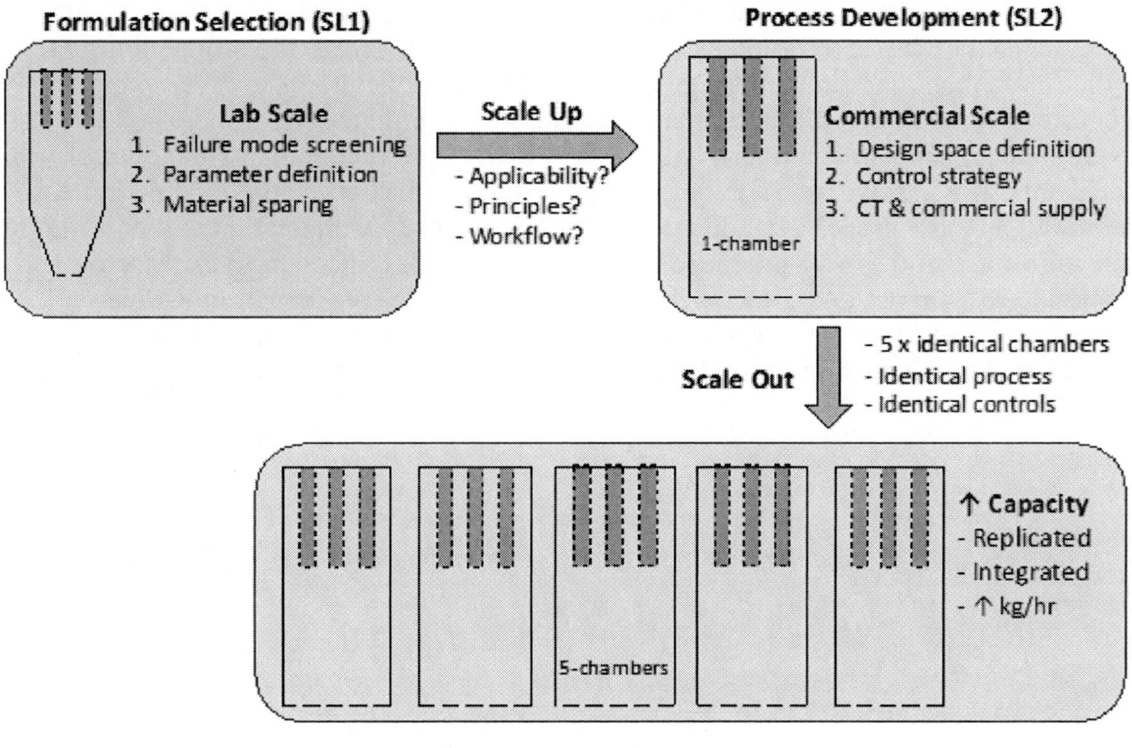

184r Antioxidants Supplementation to Cell Culture Media and Feed Influences Productivity and Product Quality of Mabs from CHO Cells

Erin Kotzbauer, Suyang Wu, Yen-An Lu, Kyle Devenney, Karen (Min-yea) Lee, Venkata S. Tayi

High expression of therapeutic proteins in Chinese Hamster Ovary (CHO) cells is crucial in biopharmaceutical processes to meet high product demand and reduce cost of goods. However, high level of recombinant protein synthesis results in high flux through metabolic and protein folding pathways which result in high oxidative stress. This oxidative stress can be characterized by high levels of reactive oxygen species (ROS), which can be detrimental to cellular function, lead to apoptosis, or can modify the desired therapeutic protein. In addition, the biopharmaceutical industry is transitioning away from the use of serum, which has antioxidant properties, and developing serum-free chemically defined media for production. To alleviate oxidative stress, cell culture media is supplemented with various antioxidants, such as glutathione, lipoic acid, retinyl acetate, ascorbic acid, and rosmarinic acid. However, some antioxidants have solubility issues in media, are toxic at high levels, and may interact with each other through similar redox pathways. Determining which antioxidants and their optimal levels in CHO cell media is key to improving cell viability and productivity for bioprocesses.

Results from this study showed that antioxidants supplemented to the production feed increased cell-specific productivity by 1.2-fold. However, antioxidants supplemented to both production media and production feed increased cell-specific productivity by 1.5-fold while inhibiting the cell growth, resulting in lower peak cell density compared to the control. To prevent early growth inhibition, antioxidants were added to the feed for intensified fed-batch reactors and to the production stage perfusion media for continuous manufacturing. This strategy enabled cultures to grow to high cell density first before introduction to the antioxidants that improved productivity. Intensified fed-batch reactors with antioxidant supplemented feed had 1.2-fold increase in titer compared to control, while perfusion bioreactors with antioxidant supplemented perfusion media had 1.3-to-1.7-fold increase in permeate titer throughout the process. In addition, the antioxidant supplementation to media and feed affected quality attributes of the mAb produced, such as increased fragmentation, increased aggregation, and lower mannose and galactosylation. This study provides insight into which antioxidants improve CHO cell culture productivity and/or viability and the strategies of supplementing these antioxidants for optimal effects.

184s Diffusion of Viral Vectors Informs Continuous Manufacturing with Aqueous Two-Phase Extraction

Seth Kriz, Sydney Dankert, Idris Tohidian, Caryn Heldt

Continuous biomanufacturing is just around the corner but many traditional unit operations are struggling to adapt quickly. That is why we recently developed a multi-step continuous purification platform for viral vectors using aqueous two-phase systems (ATPS). Recoveries of up to 80% of infectious titer, multiple log removal of host cell protein, and consistent performance among enveloped and non-enveloped viruses make it a robust tool for vaccine or gene therapy manufacturing. An economic analysis of our ATPS-based process shows a 90% reduction in capital costs and 50% lower operating cost compared to traditional chromatography-based manufacturing. During this development, a strong dependence on mass transport related to mixing and settling was observed. To speed optimization for new viral vectors, parameters describing this separation process are needed. In this work, diffusion coefficients in each aqueous phase and mass flux across the ATPS interface are measured for protein and virus to determine the contributions of diffusion during the separation. A Y-channel microdevice allows capture of real-time diffusion using fluorescently labeled particles and comparison with calculated models in COMSOL Multiphysics software is used to identify transport parameters. These results will inform scale up of ATPS purification steps as we continue toward demonstrating fully continuous end-to-end viral vector manufacturing.

184t Unlock Bioprocessing Capacity: A Comparative Study Focused on Intensified Downstream Processes

Xuan Han, Avani Ramachandra, Dan McNamara, Paul Gramlich

In response to the swiftly evolving biologics landscape marked by a rapid development cycle and expansive market, the biopharmaceutical industry faces an imperative need to optimize manufacturing process for faster, cost-effective production. This study presents a comparative analysis investigating the downstream processing capability for higher-productivity cell culture harvest by assessing three distinctive purification trains. Among these, two adhered to widely-adopted approaches combining bind-and-elute chromatography with a flowthrough polishing step, while the third employed an intensified process with two flowthrough columns. All three processes effectively purified the product of interest from feed material, even in the presence of elevated aggregation. While subtle differences were observed for impurity removal, the attributes monitored during in-process steps remained consistent across the three processes. In addition, the study evaluated process modeling to further compare process performance. Notably, the intensified process outperforms the other two in terms of throughput and operational efficiency. These results underscore the need to integrate intensified downstream processing technologies in antibody production. These results offer a comprehensive comparative analysis shedding light on opportunities and challenges in contemporary intensified bioprocessing

184u Machine Learning and Artificial Intelligence for Formulation Optimization

Howard Stamato, Daniela Blanco

Formulation of oral solids is a challenging topic because of the difficulty of dealing with powder properties alone. The number of materials available for use, including the variations on their manufacture and combination further complicates the task. When formulating a new molecular entity, the most important ingredient is the least well characterized. Additional considerations for the stability, release rate, size, shape, swallowability, identifying markings or even ability to split the tablet can be added to the amount of information to be processed.

While with modern techniques of statistically designed experiments, material property characterization, advanced analytical technology the data is more meaningful, the size of the dataset and the number of factors to consider only grows. Even after a lifetime of experience a formulation team often resorts to analog neural nets of what worked before, its command of the graphs they have absorbed from the literature, and the limited experimentation they have resources for.

Recent advances in computing power have fueled new capabilities in machine learning (ML) and artificial intelligence (AI). The digital neural net can process more factors in minutes than the analog net is capable of after years of training. New insights from the ML/AI approach provide a faster path to selection, optimization, and better use resources in implementing formulation designs.

This work describes a collaboration leveraging prior knowledge of formulation design and SuntheticsML, a user-friendly ML platform, to leverage a training set of data from a matrix controlled release oral solid dosage form. SuntheticsML is an accessible online ML tool tailored for researchers without coding or ML expertise. The model platform was intuitive, requiring only an introduction to the major features and some practical use to learn. Although specialized training would be valuable in understanding all the features available, value could be derived on essentially a walk up basis from someone experienced in the basics of formulation and simple statistical designs or models.

SuntheticsML provides researchers with versatility across applications in formulation optimization, proficiently handling numeric, discrete, and mixed-integer optimization problems with up to 20 input parameters. This ML-powered approach allows flexible execution in serial or parallel experimentation, accommodating any user-specified budget. Furthermore, it facilitates bounded-target, multi-objective, and constrained-input optimizations, enabling simultaneous enhancements in cost and material efficiency.

The composition was studied across two actives of similar properties and in two regions of release rate of interest. The space encompassed 8 excipients, and multiple dosage form presentations. The training set was 19 laboratory scale (few hundred gram to 1 kg scale) experiments. The experimental data was at hand from different sources including small designed studies or initial efforts at formulation. The Sunthetics ML/AI platform for composition and process data science was used to create an ML/AI model to generate insights and predictions from the training set and establish the design space.

Use of the ML/AI model enabled finding the right operating region to achieve the tablet performance and optimized that performance to a goal provided by the clinical team more quickly. Specifically, the platform showed that several different grades of excipients could be used to achieve comparable results opening the possibility to supply chain and formulation preferences to achieve goals. A key value of the platform is the built-in visualization tools. The visualization toolbox allows the team to look at the data from several perspectives thus assembling an understanding of multiple variable interactions gaining new insights into the drivers for critical quality attributes. A slide bar function gives immediate feedback to likely outcome of proposed formulation and can aid team discussion when deliberating options. The model also gives several types of suggested optimal points and suggested experiments to improve the reliability of the dataset.

Use of the model eliminated several iterations of designed experiments and led to the identification of a formulation meeting the target profile on an expedited timeline.

184v Data 1.0 before Pharma 4.0: From RandD to Supply Chain

Amy Wagner, Nima Yazdanpanah

The fundamental building block for AI, ML, Simulation, Pharma 4.0, and other sophisticated approaches is the DATA and the importance of its quality and availability (i.e. accessibility, cleanness, correctness, comprehensiveness, context, ..). However, this fundamental aspect mostly ignored, either the data is in silos, trapped in dispersed types and ELNs and vendor licensed platform, messy, noisy, incomplete, and so on. Although there have been different efforts in harmonizing the format, connectivity, and accessibility, still a sharable and customizable database and a collaborative platform (either for data governance or running models and simulation) is a big challenge.

In this presentation we will discuss on the complexity of data in the pharmaceutical industry from R&D to post manufacturing and demonstrate some work arounds and solutions for data curation and running models across organizations. The case studies will be from different types of applications and organization (such as CDMs, CMOs, biologics, small molecule, and supply chain). Democratization of numerical models and sharing models across stakeholders will be demonstrated on Cloud based and user friendly interfaces.

184w Federated Hierarchical Tensor Networks: A Collaborative Learning Quantum AI-Driven Framework for Healthcare

Amandeep Bhatia, David E. Bernal Neira

Healthcare industries frequently handle sensitive and proprietary data, and due to strict privacy regulations, they are often reluctant to share data directly. In today's context, Federated Learning (FL) stands out as a crucial remedy, facilitating the rapid advancement of distributed machine learning while effectively managing critical concerns regarding data privacy and governance. The fusion of federated learning and quantum computing represents a groundbreaking interdisciplinary approach with immense potential for revolutionizing various industries ranging from healthcare to finance. In this work, we proposed a federated learning framework based on quantum tensor networks, which leverages the principles of many-body quantum physics. There are currently no known classical tensor networks implemented in federated settings. Furthermore, we investigated the effectiveness and feasibility of the proposed framework by conducting a differential privacy analysis to ensure the security of sensitive data across healthcare institutions. Experiments on popular medical image datasets show that the federated quantum tensor network model achieved a mean receiver-operator characteristic area-under-the-curve (ROC-AUC) between 0.91-0.98. Experimental results demonstrate that the quantum federated global model, consisting of highly entangled tensor network structures, showed better generalization and robustness and achieved higher testing accuracy, surpassing the performance of locally trained clients under unbalanced data distributions among healthcare institutions.

Advancing medical research through collaborative healthcare data sharing presents challenges stemming from data heterogeneity (in terms of formats, standards, and structures) and stringent privacy regulations. The concept of federated learning (FL) holds great promise for advancing medical research by enabling collaborative analysis of decentralized data and addressing challenges related to data heterogeneity, communication, security, technical complexity, and legal considerations. Initially, FL was designed for various domains, including edge and mobile device use cases [1], but it has recently gained momentum in the realm of healthcare applications. It allows for model training on edge nodes, enabling real-time insights and decision-making without the need for extensive data transfers to a central server.

In the past few years, quantum computing, especially distributed quantum computing, including quantum machine learning (QML), has made remarkable progress. Its extraordinary ability to leverage the combined power of distributed quantum resources surpasses the constraints of individual quantum nodes [2]. In parallel, Tensor networks (TNs) are considered promising candidates for quantum machine learning (QML) architectures [3].

Considering that tensor networks can serve as representations for both neural networks and quantum circuits [4], it becomes a natural objective to investigate the intersection of these two fields in federated settings. The main motivation is to unlock the potential benefits of quantum tensor networks and federated learning in the healthcare sector. This includes creating improved diagnostic tools, enhancements in medical image analysis, and ensuring meaningful results for rare diseases when smaller healthcare institutions lack sufficient data to train an accurate predictive model.

In this paper, we proposed a quantum tensor networks-based federated learning framework, or FedQTN, for collaborative learning between multiple healthcare institutions, enabling QTNs to train on non-independent and identically distributed medical images. To validate our approach and its adaptability to heterogeneous data, we demonstrated the learning capability of FedQTNs on popular medical image classification datasets, including chest radiographs from the RSNA and NIH datasets, MRI brain scans from the Alzheimer's ADNI dataset, and CT-scans from the CT-kidney dataset.

Keywords: quantum machine learning, federated learning, distributed optimization, healthcare, medical imaging, classification, privacy-preserving

References

1. McMahan, B., Moore, E., Ramage, D., Hampson, S. & Arcas, B. Communication-efficient learning of deep networks from decentralized data. Artificial Intelligence And Statistics. pp. 1273-1282 (2017)

2. Cuomo, D., Caleffi, M., & Cacciapuoti, A. S. (2020). Towards a distributed quantum computing ecosystem. IET Quantum Communication, 1(1), 3-8.

3. Rieser, H. M., Köster, F., & Raulf, A. P. (2023). Tensor networks for quantum machine learning. Proceedings of the Royal Society A, 479(2275), 20230218.
4. Grant, E., Benedetti, M., Cao, S., Hallam, A., Lockhart, J., Stojevic, V., & Severini, S. (2018). Hierarchical quantum classifiers. npj Quantum Information, 4(1), 65.

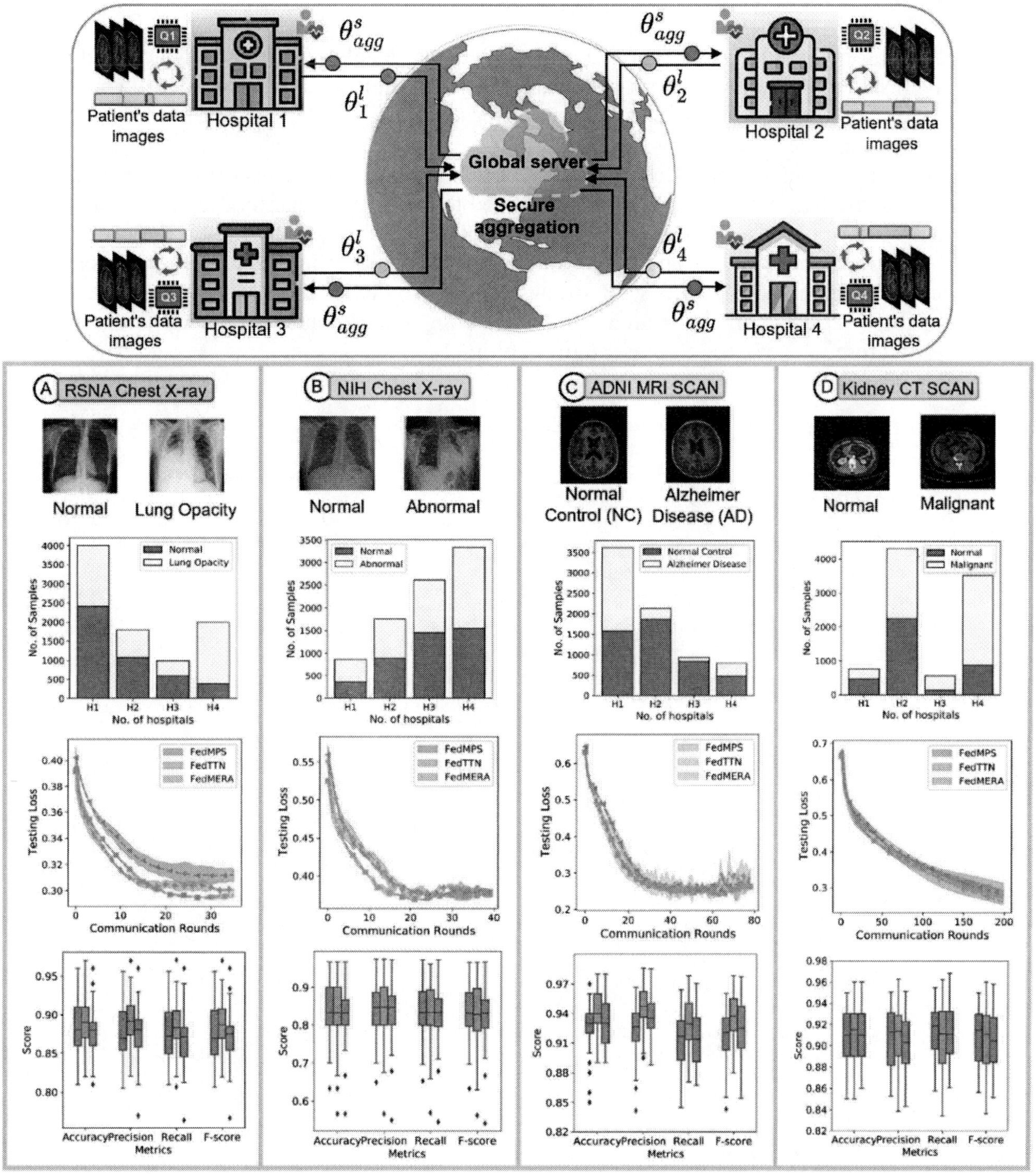

184x Hybrid Modelling for Semi-Permeable Membranes: A Water Vapor Transmission Rate (WVTR) Stability Estimation.

Mandakini Padhi, Nallasivam Angannan, Jean Sebastien Simard

Stability studies for sterile injectables are an essential step in developing a globally acceptable drug product. The products presented in semi-permeable polymer films tend to lose water when exposed to the standard temperature & relative humidity conditions, following the ICHQ1A guideline. The proposed hybrid model utilizes a constant flux, quasi-steady state approach to estimate the % water loss, using the first-principle permeability model. The existing data from historical cases and extended predictions from the first-principle model were utilized to develop a Machine Learning (ML) based user interface (UI), specific to different types of semi-permeable packaging material currently utilized by the Pfizer community. The UI can be fully leveraged to estimate the water loss % at a longer duration, and substantial long-term WVTR studies can be avoided or estimated well ahead. The model is mathematically validated and compared with the existing experimental data, and the maximum error was found to be < 2.5% for all presentation volumes.

184z Spray Dryer Design for Robust Manufacturing of Amorphous Solid Dispersions

Gabriele Sadowski, Christian Lübbert

An established enabling technique for enhancing the bioavailability of poorly soluble drugs are ASDs (amorphous solid dispersions), in which the drug molecule is amorphously embedded in a suitable polymer matrix. Spray drying is one of the major techniques for manufacturing ASDs with high melting drugs. The development of an appropriate and robust spray-drying process is challenging as many different variables such as spray drying process parameters or solvents must be selected carefully and tailored to individual needs of the polymer and API. The selected solvent must dissolve enough drug and polymer to generate the spray dryer feed solution. Solvent mixtures can help dissolving cellulose-based polymers such as HPMCAS which are not soluble in single solvents. These solvent mixtures tend reveal different evaporation velocities due to their differing volatilities and boiling points and might thus unevenly evaporate during the droplets' drying process, leading to a temporary accumulation of one of the solvents. Additionally, the solvent can induce unwanted and hardly detectable amorphous phase separation in the particles during the drying progress, even at the extreme evaporation rates within a spray dryer. The finally obtained spray-dried material always contains residual solvent which usually requires secondary solvent removal. The presentation will highlight the revolutionary potential of using in-silico predictions to address the above-mentioned challenges, from solvent/ solvent mixture screening to predicting the accumulation risk and residual solvent content. This approach using the thermodynamic model PC-SAFT allows predicting scale-up behavior, process design spaces and perform material-property specific sensitivity optimization of process parameters. It requires little experimental training data sets that are typically available at early stages of formulation development. The formulator gains fundamental understanding about the intermolecular interactions within a formulation from a computational in-silico perspective. Development budgets are spent more economically by substituting intensive and misleading trial-and-error screenings or intensive design of experiment studies by this in-silico technique.

184aa Advancing Kinetic Parameter Estimation: A Metabolism-Informed Variational Bayesian Inference Approach

Mariana Monteiro, Ranjith Chiplunkar, Max Mowbray, Cleo Kontoravdi

The biopharmaceutical industry must ensure safe and consistent biological products. To meet these demands, mathematical models are increasingly used to design, predict, optimize, and control processes, from development to manufacturing. In the field of mammalian cell cultures, which accounts for 80% of the commercially available therapeutic proteins [1], the exploitation of mathematical models, such as kinetic models, has also become a useful tool to maximize information gained from experimentation and to reduce expense. However, developing reliable kinetic models for cell cultures is challenging as they are typically parameterized for a specific range of conditions and often face challenges related to generalizability. This leads to limited extrapolation power of the model and a lack of trust in the estimated parameters, which restricts the model exploitation.

One of the common ways of addressing the extrapolation power of models is through the integration of different models and their philosophies. While, ideally, mechanistic models such as the flux balance analysis (FBA) approach alone should be able to produce reliable results, these are often limited either by a poor understanding of mechanisms, or the simplistic assumptions that are made to facilitate their solution through partial representation of biological pathways. Hence, hybrid modeling approaches that combine different modeling paradigms resulting in the best of both worlds are often preferred. An example of this is provided by dynamic FBA, which integrates common semi-empirical kinetic model structures and FBA [2].

In this work, we propose a hybrid model construction framework that leverages variational Bayesian inference (VBI) for the parameterisation of a kinetic cell culture model informed by a stoichiometric metabolic model. VBI allows one to conveniently combine different modelling paradigms along with the experimental data to obtain probabilistic hybrid models. Unlike traditional methods such as maximum likelihood estimation, it allows the embedding of mechanistic knowledge via the prior distribution. Our model uses metabolic knowledge from flux sampling (FS) and FBA as priors for the estimation of the model's parameters. By using distributions from FS and FBA as prior knowledge, our framework can utilize mechanistic knowledge from a metabolic network model, thus improving the reliability of the kinetic parameters and generalizability of the model. Moreover, being a probabilistic model, it allows for the quantification of model uncertainty through the distribution of the parameters while also handling measurement uncertainties.

We demonstrate the applicability of our framework in the estimation of kinetic parameters of a Chinese Hamster Ovary cell culture model, using published data from a fed-batch experiment [3]. Available daily measurements include viable cell density, viability and extracellular glucose, lactate, ammonia, and amino acid concentrations. The priors for the yields are obtained via FS a metabolic model constrained by experimental data and those for maximum growth rate are obtained by solving a metabolic model with FBA under a biomass maximization objective. The priors for the remaining parameters are given through the analysis of metabolite concentrations. Finally, the posterior distributions of the model parameters are estimated using the coordinate ascent mean-field variational inference approach [4]. Our methodology successfully estimates kinetic parameters, and additionally predicts asparagine concentration, a typically unmeasured growth-limiting metabolite.

The incorporation of mechanistic knowledge in the proposed framework improves the generalizability of the obtained models rendering them suitable for optimization and control of cell cultures over a wide range of operating conditions. Additionally, the framework also allows the models to be conveniently adapted to new experimental data, showcasing the versatility of the proposed framework.

References:

[1] Al-Majmaie, R., Kuystermans, D., & Al-Rubeai, M. (2022). Biopharmaceuticals produced from cultivated mammalian cells. *In Cell Culture Engineering and Technology: In appreciation to Professor Mohamed Al-Rubeai* (pp. 3-52). Cham: Springer International Publishing.

[2] Mahadevan, R., Edwards, J. S., & Doyle, F. J. (2002). Dynamic flux balance analysis of diauxic growth in Escherichia coli. *Biophysical journal*, 83(3), 1331-1340.

[3] Kyriakopoulos, S., & Kontoravdi, C. (2014). A framework for the systematic design of fed-batch strategies in mammalian cell culture. *Biotechnology and Bioengineering*, 111(12), 2466-2476.

[4] Blei, D. M., Kucukelbir, A., & McAuliffe, J. D. (2017). Variational inference: A review for statisticians. *Journal of the American statistical Association*, 112(518), 859-877.

184ab Efficient Multi-Objective Bayesian Optimization of Chemical Reactions

Vidhyadhar Manee, Thomas Tampone, Dauzhan Yerkozhanov, Zhen Lei, Frederic Buono

Chemical Reaction Optimization in Early-Stage Drug Development typically involves two stages. In the first stage, the objective(s) of the optimization campaign are clearly formulated and the design parameters that have a measurable impact on the reaction are identified. Stage 2 involves running experiments within the identified design space to discover the optimal conditions that maximize the objectives. Even with the adoption of automated reaction platforms[1] and parallel experimentation, a thorough exploration of the design space is impractical for most real-world problems and the complexity scales with the dimensionality of the design space. Machine Learning algorithms are useful in this context since they can guide the sequence of experiments to be run by making informed data-driven decisions. Bayesian Optimization (BO) has been the algorithm of choice in recent data-driven reaction optimization literature due to its ability to find the optimal reaction conditions while simultaneously satisfying time and material constraints. In the simplest case where a single objective has to be maximized, BO works by fitting a probabilistic surrogate model that maps the design parameters to the objective. Based on the predictions of the learned surrogate model and the associated uncertainties, the experiment that is likely to improve upon the objective is carried out.

Often, there is more than one objective of interest to the chemist and these objectives can compete with each other (For example, Yield, Space-Time-Yield, E-factor, cost-of-materials etc.). Here, a multi-objective BO (MOBO) algorithm is preferred where independent surrogate models are built for each objective. The goal in this scenario is not to maximize any one objective but to discover the set of pareto-optimal reaction conditions. In this work, we identify two inefficiencies associated with how MOBO algorithms are set up and propose solutions that can improve the efficiency with which they discover the optimal reaction conditions.

Current MOBO algorithms treat each objective as independent unknown black-box functions. In practice, this is often not the case. As a trivial example, if cost-of-materials is one of the objectives that needs to be optimized, a surrogate model of cost-of-materials does not need to be learnt from data since the mapping between the design parameters and the cost-of-materials can be calculated beforehand. While most objectives are not this trivial, many are often not completely black box in nature either. For example, Space-Time-Yield, which is a common objective of interest in MOBO, can be expressed as the composition of an unknown black-box function (yield) and known design parameters. By treating objectives as grey-box instead of black-box in this way, Raul Astudillo et al.[2] have shown that single-objective BO algorithms can significantly improve their query-efficiency. Inspired by this approach, we extend this idea to the multi-objective scenario and show improvements in time and material requirements.

Another limitation with current MOBO algorithms is that all objectives are weighted equally, and the algorithm is incentivized to find the trade-off between these objectives (pareto front). In reality, some objectives are more important to the chemist than others (e.g. Yield over E-factor) resulting in only a portion of the pareto front being relevant. This leads to the algorithm expending its resources wastefully on discovering uninteresting conditions. Lin et al.[3] have recently shown that this problem can be alleviated by having a surrogate model learn the preferences of the expert decision maker through pairwise comparisons. After learning a preference surrogate model, the BO algorithm is tasked with optimizing the expert's preferences. It is not clear at the outset if this approach can transfer over to the reaction optimization problem since a chemist's preferences have never been learnt through pairwise comparisons. In the second part of this work, we model the preferences of a chemist using a preference based Gaussian Process model and compare this approach of preference-based BO to conventional MOBO.

1. Sagmeister, P., Ort, F. F., Jusner, C. E., Hebrault, D., Tampone, T., Buono, F. G., Williams, J. D., & Kappe, C. O. (2022). Autonomous Multi-Step and Multi-Objective Optimization Facilitated by Real-Time Process Analytics. *Advanced Science*, *9*(10), 2105547. https://doi.org/10.1002/advs.202105547
2. Astudillo, R., & Frazier, P. I. (2019). Bayesian Optimization of Composite Functions. *ArXiv*. https://doi.org/10.48550/arxiv.1906.01537
3. Lin, Z. J., Astudillo, R., Frazier, P. I., & Bakshy, E. (2022). Preference Exploration for Efficient Bayesian Optimization with Multiple Outcomes. *ArXiv*. https://doi.org/10.48550/arxiv.2203.11382

184ac Scaling Organic Electrosynthesis: The Crucial Interplay between Mechanism and Mass Transport

Zachary Oliver, Dylan Abrams, Luana Cardinale, Greg Beutner, Seb Caille, Benjamin M. Cohen, Lin Deng, Moiz Diwan, Antonio Ferretti, Michael Frederick, Kaid Harper, Dan Lehnherr, Tessa Myren, Kyle Quasdorf, Melda Sezen-Edmonds, Kevin Stone, Shannon S. Stahl, Marcel Schreier

Recent progress in the field of organic electrosynthesis shows promise to develop chemistries of pharmaceutical interest that utilize the electron as a "green" reagent. However, the advances made in organic electrosynthesis have not yet been met by equal progress in their implementation at larger scale, hampering their use in the production of active pharmaceutical ingredients (APIs). Presently, most organic electrosynthesis reactions are studied in simple parallel plate or batch cells using small surface area electrodes. The flow profile and mass transport behavior of these cells, however, does not generally make them appropriate for synthesis at scale. Over the past century, significant developments in electrochemical reactor engineering have led to improved reactor designs, but these have found only limited application to modern organic electrosynthesis.

In this talk, I will discuss the performance of advanced electrochemical reactor designs towards contemporary organic electrosynthesis. Our results demonstrate that gaining independent control over the flow profile and mass transport in electrochemical reactors provides a novel, and thus far unexplored avenue for controlling the selectivity observed in electrosynthesis reactions. Using optimized reactor designs and reaction conditions, we reach unprecedented selectivity and rates for organic electrosynthesis, approaching the multi-kg per day rates necessary for early-stage clinical trials.

Our work provides a generalizable framework for understanding the interaction between fundamental reaction kinetics, reactor geometry, and mass transport to control the rate and selectivity of pharmaceutical organic electrosynthesis at scale.

184ad Toxic to Treatment: Bioprospecting in Battery Waste Rhizospheres for Novel Therapeutics Against High Priority Antibiotic-Resistant Pathogens

Kylah Dalgleish, John Unuofin, Luke Invernizzi, Michael Daramola, Samuel Iwarere

The rate of antibiotic resistance on the rise has urgently called for a radical approach to developing sustainable solutions to mitigate it. In particular, two critical pathogens, *Pseudomonas aeruginosa* (strain ATCC 27853), and *Acinetobacter baumannii* (strain ATCC-BAA-1605), have been classified as extremely drug-resistant (XDR) by the World Health Organization (WHO), subsequently overwhelming contemporary treatments. Consequently, this study hypothesised that rhizospheric microbiota of plants tolerant to metal-laden environments must have demonstrated resilience through their survival under nutrient limiting conditions and extreme environmental stress, and therefore might secrete antimicrobials during competitive exclusion that are potent enough to inhibit the growth of these pathogens. To investigate our hypothesis, this study evaluates the secretion of strong antimicrobials from heavy-metal-contaminated rhizospheres using a novel method that induced unnatural environmental stress on them. The success of this method was indicated by the experiential inhibition (≤ 19 mm) of 3.0 × 10^14 CFU/mL estimates of both multi-drug-resistant pathogens, during Agar Well Diffusion tests using crude extracts of the secondary metabolites. Moreover, pathogens were more responsive to exudates of microbiota from environmentally stressed rhizospheres than their counterparts from an organic rhizosphere. The components of the extracted secondary metabolites were purified and separated using a generic fractionation using HPLC-MS/UV, where a number of compounds demonstrated activity against the pathogens in the form of halos of inhibition. Ultimately, the novelty of the metabolites secreted needs to be validated through UPLC-MS and NMR, and the IC50 needs to be calculated to warrant further investigations on their pivotal contributions to biomedicine and biopharmaceutics.

Keywords: Antimicrobial; Antibiotics resistance; Antibiotics susceptibility; Competitive exclusion; Environmental stress; Rhizospheric microorganisms

151a Towards Sustainable Biopharmaceutical Production Via Single-Use Technologies

Behnam Partopour

Advanced therapeutics such as monoclonal antibodies (mAbs) are among the most powerful tools for improving global health. These proteins, from rare cancers to viral infections, enable us to fight the most challenging diseases. Therefore, a more equitable production and distribution of mAbs is essential for global health. As demand for these therapeutics increases, so does the environmental impact of their manufacturing processes. Today, addressing these ecological impacts has become a top priority of technology providers and biopharmaceutical manufacturers.

Single-use technologies generate significant opportunities for process intensification and more robust and cost-effective biopharmaceutical manufacturing which can result in reduced facility footprint and energy consumption. However, the increasing amount of consumable waste produced in the sector is concerning.

In this talk, we review the potential environmental impacts of mAbs biomanufacturing and quantify how different process intensification strategies contribute to more sustainable production. Furthermore, you will hear about Sartorius's contribution towards "Sustainability Development Goals", "New Plastic Economy", and examples of circularity opportunities, potentially, applicable to the biopharmaceutical industry.

151b Invited Talk: Engineering a Sustainable Future: Cultivating a Culture of Innovation and Environmental Stewardship

Margaret Faul

In the pursuit of a sustainable future, the role of scientists and engineers, working in cross functional teams, is pivotal in driving innovation and fostering environmental stewardship across industries. This presentation explores how a culture of sustainability can be effectively cultivated within teams, emphasizing the integration of green practices into every aspect of the product lifecycle. By showcasing successful case studies and the application of cutting-edge biomanufacturing technologies, we will highlight the impact of cross-functional collaboration and the commitment to continuous improvement. Attendees will gain insights into actionable strategies for embedding sustainability into their operations, enabling their organizations to not only meet ambitious environmental targets but also lead the way in the global effort to protect our planet for future generations.

About the speaker:

Margaret received her Ph.D. degree in Synthetic Organic Chemistry from Harvard University. She is currently Vice President of Drug Substance Technologies and Amgen Massachusetts Site head.

Margaret joined Amgen in 2003 in the Small Molecule Drug Substance Organization and has held numerous positions of increasing responsibility including Executive Director of the Pivotal Drug Substance Technologies, Vice President of Drug Product Technologies, and Vice President of the Manufacturing and Clinical Supply Chain organization. She has supported the advancement of the Amgen portfolio from clinical through to commercial development and been instrumental in the integration of innovative technologies into the manufacturing network. Margaret has invested a significant effort in evolving a green and sustainable culture in the workplace and these efforts have been recognized by Amgen being a recipient of the Presidential Green Chemistry Challenge Award.

Margaret has a strong external scientific presence with >100 publications in peer review journals. She is a Fellow of the American Chemical Society and an elected member of the National Academy of Engineering. She is a past chair of the Board of Directors of the International Consortium for Innovation and Quality in the Pharmaceutical Industry and a founding member of the Enabling Technologies Consortium. Margaret is an associate editor for Organic Letters and is a member of the editorial boards for Science of Synthesis, Organic Syntheses, Synthesis and SynLett, and Advanced Synthesis and Catalysis. In partnership with Thiéme and the Science of Synthesis she sponsors the Dr. Margaret Faul Award for Women in the Chemical Sciences.

151c Improving Environmental Sustainability in Medical Devices: Strategies and Opportunities at Lilly

Zeming Gou, Markus Wierzoch

Markus Wierzoch and Zeming Gou will discuss sustainability in the medical device sector, emphasizing the importance and challenges of integrating planet-centric practices, while focusing on improving health outcomes through human-centered design.

The presentation will discuss Lilly's 2030 Environmental Goals and outline the strategies and opportunities the company has identified to meet its sustainability objectives regarding delivery devices.

Key focus areas will include the adoption of renewable materials, innovative sustainable packaging solutions, and takeback and recycling programs.

The presentation will center around a case study on Lilly's drug delivery devices, which will illustrate the application of life cycle assessments (LCA) in evaluating and improving their environmental impact.

This may provide a deeper understanding of how LCAs can inform and drive sustainability efforts in product design and development.

150a Regulatory Considerations for Antibody-Drug Conjugates (ADCs) (invited talk)

Kanny Wan

Antibody-drug conjugates (ADCs) are amongst the fastest growing drug classes in oncology, and their use could expand to other therapeutic areas in the future. ADCs are highly complicated molecules, that include a combination of a monoclonal antibody, a cleavable or non-cleavable linker, and a drug payload. Development of ADCs is challenging because ADCs are very complex combination products for which there is minimal regulatory guidance.

The rapid increasing trend of novel ADC products includes a variety of antibody-drug linker conjugation technologies, drug payloads, linkers, and mechanisms of action creating new challenges in the development, manufacturing, and controls of these increasingly complex products.

The presentation will focus on the following points:

- FDA approved ADCs and new trends in payload drugs, linkers, and conjugation reactions.
- Overview of the Food and Drug Administration (FDA) review process for ADC products.
- Regulatory expectations on payload-linker, ADC drug substance, and ADC drug product controls.

150b High Throughput Platform for Macromolecular Transport Screening

Paulina Babiak, Chenji Li, Carly Battistoni, Mazin Hakim, Adib Ahmadzadegan, Alyssa Panitch, Arezoo Ardekani, Pavlos Vlachos, Luis Solorio, Julie Liu

The mass transport of a therapeutic injected into the subcutaneous tissue is dictated by physiochemical properties of the molecule such as size and electrostatic charge. Bioavailability and efficacy of the therapeutic formulation depend on efficient transport of the molecule from the injection site to lymphatic or blood vessels. In vitro transport screening platforms provide insights into the effects of tissue and therapeutic properties on macromolecular transport through biological barriers. In this work, we focus on Transwell macromolecular recovery platform, which is an economical and high-throughput method, to systematically evaluate effects of extracellular matrix components (e.g. collagen types I, II and III, and hyaluronic acid (HA)) on mass transport properties of model macromolecules. Transport of macromolecules which are representative of current therapeutics in subcutaneous injections was examined.

We demonstrate that macromolecular transport within collagen hydrogels is inversely related to the hydrodynamic radius of the diffusing macromolecules. Blended collagen hydrogels containing collagen type I and II, or I and III had altered fibril morphologies (smaller fibril formation) compared to pure collagen type I hydrogels. The smaller fibrils of blended collagen hydrogels resulted in decreased molecular mass recovery rate compared to pure collagen type I gels. Transport through combined collagen type I and HA (ColHA) hydrogels was also assessed. Increasing HA concentration within the ColHA hydrogels decreased macromolecular recovery due to increased viscosity within the matrix. The inverse relationship between hydrodynamic radius of the diffusing molecules and recovery rates persisted in both blended collagen and ColHA hydrogels.

Recovery rates of large molecules such as immunoglobulin G (IgG), a molecule similar in size to therapeutic antibodies, were highly sensitive to HA concentration in col hydrogels. Smaller molecules, such as myoglobin and lysozyme, that are similar in size to insulin, experience electrostatic effects as HA concentration increases within collagen gels. Recovery of macromolecules in HA solution was a function of both electrostatic and steric interactions. Our results demonstrate that the Transwell platform can be utilized for systematic evaluation of therapeutic transport as a function of molecular characteristics. The results presented can inform biotherapeutic physiochemical properties for efficient transport within the subcutaneous tissue.

150c Investigate the Druggability of Protein Targets for *Listeria Monocytogenes* for Drug Discovery

Robert Hanes Jr., Zuyi Huang

It is reported that approximately 48 million cases of foodborne illnesses take place in the U.S. every year. Among them are approximately 1600 cases of listeriosis that result in about 260 deaths. Vulnerable populations, including the elderly, children, and pregnant women, face a fatality rate as high as 30%. Listeriosis, caused by *Listeria monocytogenes*, ranks as the third leading cause of death due to foodborne pathogens. *L. monocytogenes*, a resilient facultative intracellular pathogen, is prevalent worldwide and poses significant challenges to the food industry due to its presence in various environments and its transmission via contaminated foods and surfaces. The primary treatment for listeriosis involves ampicillin, yet concerns persist regarding the emergence of antibiotic resistance due to lateral gene transfer.

Efforts to control *L. monocytogenes* focus on disrupting stress response, virulence, and antibiotic resistance mechanisms. Enhancements in these areas could improve food preservation techniques, diminish virulence, and counteract antibiotic resistance, thereby positively impacting *L. monocytogenes* treatment. Developing novel treatments presents challenges, with Phase II drug failures often attributed to lack of efficacy. Effective drug targets should play a functional role in the disease process and be druggable. However, little work has been done to evaluate the druggability of potential targets for *L. monocytogenes*. This study aims to address this gap by developing a druggability-evaluation model based on protein binding pocket structure features. In particular, a ligand-protein computational platform was implemented to extract structural features of 535 binding pockets for known antimicrobials. These features were analyzed to identify the key features that can be used as the input of a logistic regression model. The pockets bound with known antimicrobials were regarded as druggable, with a value of one for druggability. On the contrary, a zero value of druggability was assigned to the pockets without binding ligands in the same protein target. The druggability is used as the output of the logic regression model. After the model was validated by existing data, it was applied to evaluate 23 key proteins from L. monocytogenes identified in existing literature. Notably, protein groEL, fliH/fliI complex, fliG, flhB, flgL, flgK, inlA, mogR, and prfA emerged as high-potential druggable targets. These findings lay the groundwork for future research aimed at identifying compounds that inhibit these druggable targets and designing experimental studies to confirm their effectiveness as drug targets.

150d Unveiling the Critical Roles of Cellular Metabolism Suppression in Antibiotic Tolerance

Sayed Golam Mohiuddin, Hanny Ngo, Mehmet Orman

Metabolic inhibitors are believed to exhibit antagonistic effects in the presence of antibiotics, as they are known to promote cell survival by inducing bacterial dormancy. However, the specific synergistic or antagonistic effects of metabolic inhibitors depend on their mechanisms of action, treatment timings, and their concentrations within treatment cultures, which require further investigation. Here, we systematically examined the synergistic and antagonistic effects of various metabolic inhibitors, such as chloramphenicol (a translation inhibitor), rifampicin (a transcription inhibitor), arsenate (an ATP production inhibitor), and thioridazine (a PMF inhibitor), in combination with ofloxacin (a quinolone antibiotic). These investigations were conducted under pre-, co-, and post-treatment conditions, encompassing a wide range of concentrations. When we applied four distinct synergy models (Highest Single Agent, Loewe, Bliss, and Zero Interaction Potency), which provide numerical scores indicating the degree of synergy (positive score), additivity (zero score), or antagonism (negative score) in drug combinations[1], we consistently observed that chloramphenicol, rifampicin, and arsenate received low or negative synergy scores, indicating a significant antagonistic relationship with ofloxacin. In contrast, thioridazine consistently received higher synergy scores, especially in pre- and co-treatment scenarios; however, its synergy decreased during post-treatment conditions. When we used multivariable linear regression analyses for all drugs and conditions tested here, a correlation between the synergy of thioridazine and its ability to suppress cellular energy metabolism became evident. While our study highlights the dual nature of metabolic inhibitors, it also underscores that compounds like PMF inhibitors can display significant potency and synergistic effects with antibiotics, suggesting their potential utility as effective anti-microbial adjuvants.

150e Design of Peptides to Inhibit Multiple *Clostridium Difficile* toxin B Subtypes in the Presence of UDP-Glucose and Mn^{2+}

Corey Febo, Carol Hall

Clostridium difficile (C. diff) infections cause diarrhea and severe colitis. *C. diff* kills 30,000 out of the 500,000 people it affects every year in the United States. Current antibiotic treatments are not always effective due to antibiotic resistance. Short peptides are being considered as potential drug candidates to prevent and treat *C. diff* infection by inhibiting the biocatalytic activity of the two toxins, TcdA and TcdB, released by the bacterium. These toxins transfer glucose from UDP-Glucose to Rho/Rac family proteins, leading to the death of colonic epithelial cells. In our group's previous efforts, a solid-phase peptide screening library and the *Pep*tide *B*inding *D*esign (PepBD) algorithm were used to design peptides to prevent the binding of UDP-Glucose to TcdB. Interestingly, one of the most effective peptides identified, SB5, was found experimentally to bind to a different site than the one where UDP-Glucose binds. Instead, this peptide bound to the site on TcdB where the Rho/Rac protein binds, thus preventing cell death by preventing the Rho/Rac family proteins from being glycosylated. We aim to determine how SB5 binds to the TcdB1 and TcdB3 subtypes when UDP-Glucose and Mn^{2+} are already bound since UDP-Glucose and Mn^{2+} are both present on TcdB in cellular environments. This may help us to design peptides with a lower binding free energy to multiple TcdB subtypes than the SB5 peptide. We have applied a computational approach that combines explicit-solvent atomistic molecular dynamics simulations and the PepBD algorithm to design new and more-effective peptides for combating *C. diff* infections.

150f Effect of Mutation on Dynamic Conformations and Druggability of KRAS Protein

Ibrahim Imam, Qing Shao

Kirsten Rat Sarcoma (KRAS) is recognized as one of the most relevant protein targets for cancer therapeutics because of its critical role in cancer development. However, the development of drugs targeting KRAS remains challenging due to a limited understanding of the effect of the mutation on the dynamic conformation of the KRAS proteins. This work will present our efforts to enhance the discovery of drugs that target KRAS proteins. We first investigate the dynamic conformation of three KRAS mutants (G12C, G12D, and G12V) using molecular dynamics simulations and accelerated molecular simulations. The focus is on revealing the mutation-induced variation in the residue-residue interaction network and the protein conformation and flexibility. We also predict the potential allosteric sites on the three KRAS mutants using an integration of the structure-aware protein language models developed in our group and MD simulations. In addition, we deploy deep learning-based molecular docking to search for potential small molecule candidates that may bind to the predicted allosteric sites of the three KRAS mutants.

150g Hierarchically Nano-Branched Particles for Sustained Adhesion and Drug Release to Bladder Tumors

JinGyun Lee, Katherine A. Trese, Taylor Ausec, Matthew Kim, Andrew Goodwin, C. Wyatt Shields IV

In recent decades, engineered nanoparticles have garnered considerable attention for drug delivery due to their high surface area-to-volume ratios, tunable drug release rates, and ability to deliver drugs to target sites deep within the body. However, weak interactions between nanoparticles and epithelia often result in poor retention times, limiting their benefits in clinical settings. To address this challenge, we developed soft microparticles with chitosan-functionalized nanostructured branches that can be loaded with a wide range of FDA-approved chemotherapeutic drugs and be extruded onto epithelial tumor sites. We prepared soft dendritic particles using poly(lactic-co-glycolic acid) (PLGA), a copolymer that is widely used in therapeutic drug delivery applications due to its well-studied biocompatibility and biodegradability (i.e., into lactic acid and glycolic acid). In highly turbulent flows, precipitation of the PLGA in an antisolvent results in the formation of numerous branched nanofibers around the particle. The large surface area due to the dendritic morphology of the particles and chitosan functionalization allowed them to strongly adhere to tumor cells and epithelial tissues, gradually release drugs, and later degrade into biocompatible products under physiological conditions. We experimentally measured the amount of chitosan molecules bound to PLGA surfaces and investigated the effect of chitosan adsorption on the surface charge of PLGA for enhanced adhesion on cancer cells. We further demonstrate that the dendritic particles display better retention on epithelial tissues such as a mouse bladders and human bladder cancer cells compared to spherical nanoparticles. Finally, we show that dendritic particles can be precisely extruded on tumor sites using a biocompatible alginate gel. While alginate gel slowly dissolves in urine, the dendritic particles remain attached on epithelia and sustainably release chemotherapeutic drugs. We demonstrate the ability of this system to encapsulate and release gemcitabine, docetaxel, and methotrexate to effectively kill bladder cancer cells *in vitro*. Our work highlights an adhesive nano-branched particle-based drug delivery system that offers high retention at target tissues for sustained drug release.

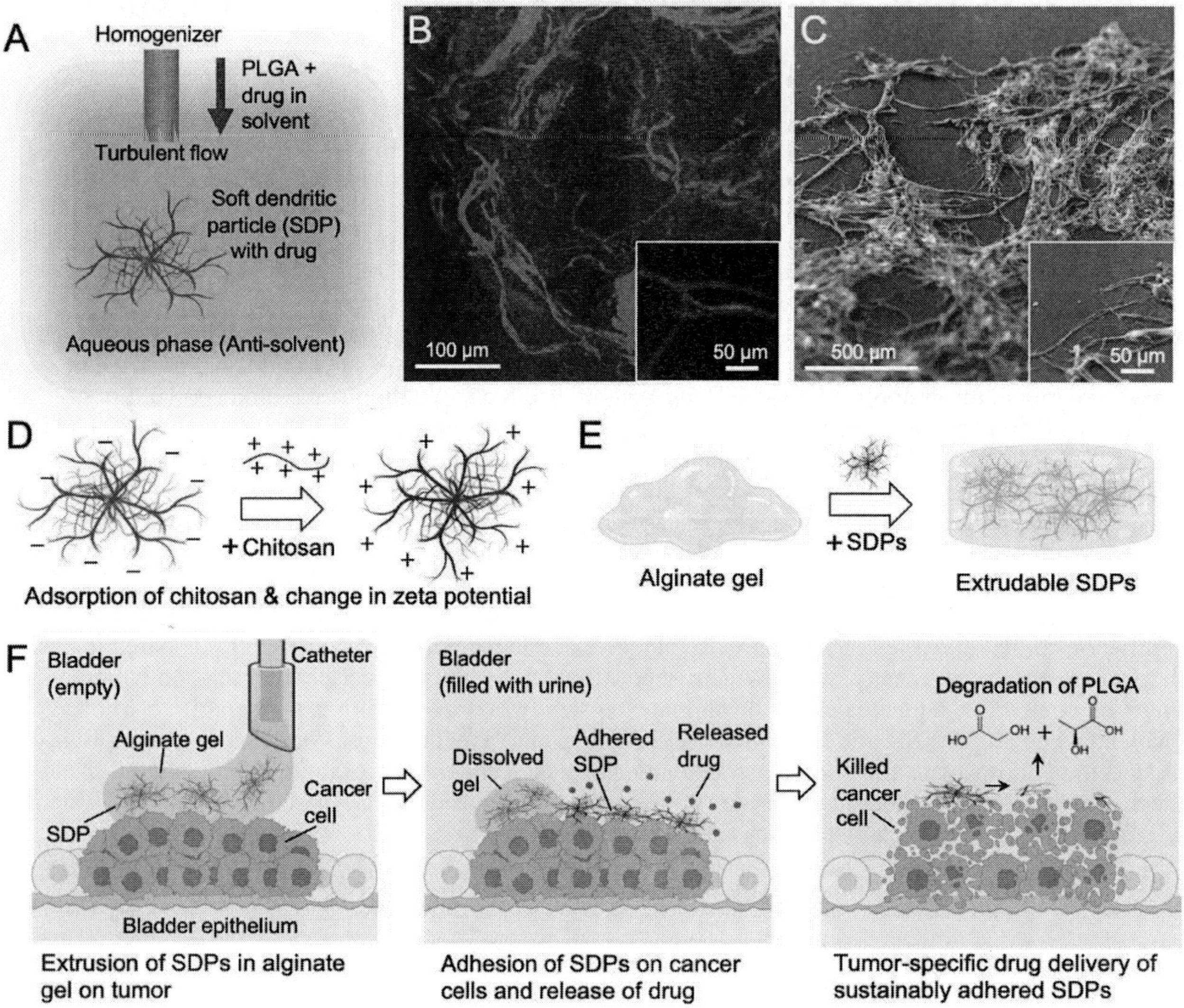
A
Homogenizer
PLGA + drug in solvent
Turbulent flow
Soft dendritic particle (SDP) with drug
Aqueous phase (Anti-solvent)
B
100 µm
50 µm
C
500 µm
50 µm
D
+ Chitosan
Adsorption of chitosan & change in zeta potential
E
Alginate gel
+ SDPs
Extrudable SDPs
F
Bladder (empty)
Catheter
Alginate gel
Cancer cell
SDP
Bladder epithelium
Extrusion of SDPs in alginate gel on tumor
Bladder (filled with urine)
Dissolved gel
Adhered SDP
Released drug
Adhesion of SDPs on cancer cells and release of drug
Degradation of PLGA
Killed cancer cell
Tumor-specific drug delivery of sustainably adhered SDPs

549a Evaluating Depth Filtration for Enzyme Removal in Small Molecule Drug Substance Manufacturing

Kevin Sirk, Bharath Krishnamurthi, Alexandra Sun, Isabella Hay

Enzyme mediated reactions are transforming the way pharmaceutical small molecules are synthesized. While enzymes have resulted in greener processes and fewer synthetic steps to make an intermediate or API, removing them from a process can be difficult. In this talk we discuss the evaluation of depth filtration as the enzyme removal unit operation for a complex, multiphase small molecule reaction system composed of immiscible organic and aqueous phases along with precipitated enzymes. Lab scale experiments were performed in a constant-flow mode to evaluate the loading capacity of different depth filter media types: positively charged cellulose fiber with diatomaceous earth (DE), positively charged polyacrylic fiber with silica, and polypropylene+cellulose+DE (Millistak+® and Clarisolve® filters). The loading capacity of filters with the same filter media but different nominal pore size ratings were also studied. Modeling was performed to determine the primary mechanism for filter fouling for all filter types. In this talk we will show that the nominal pore size of the filters had the greatest impact on loading capacity versus the type of filter media. As the nominal pore size increased, the loading capacity increased without any breakthrough. The filter fouling modeling showed that as the nominal pore size increased, the mechanism of fouling transitioned from cake formation to the standard model.

549b Spray Drying to Overcome Cold Storage of RNA-Loaded Lipid Nanoparticles

Ryan Mowry, Tanner Corrado, Gabriella Dube, Mark Kastantin

Lipid Nanoparticles (LNPs) are emerging as delivery vehicles for nucleic acid therapeutics, as evidenced by LNP-based ribonucleic acid (RNA) vaccines for COVID-19. LNPs serve as a delivery platform for encapsulating nucleic acids that are otherwise too unstable in physiological conditions to achieve the desired therapeutic effect. Even when encapsulated, chemical instability of RNA and some of the constituent lipids (over longer time scales than for RNA) necessitates cold chain handling (-20 °C and -70 °C for the Moderna and Pfizer COVID-19 vaccines, respectively) that increases cost and logistical challenges. One way to improve stability is by formulating these products in the solid-state. Solid-state formulations can slow degradation relative to liquid suspensions through two mechanisms: 1) removal of water from the formulation, which is required for many relevant degradation reaction pathways and 2) reduction in molecular mobility in the solid state relative to a liquid. Lyophilization and spray drying have been investigated for drying LNP-based products. Between the two, spray drying is more scalable, offers greater particle engineering capabilities (that may enable pulmonary or nasal delivery), and does not impose a freezing stress on the LNPs.

This talk focuses on spray drying of solid-state LNP formulations. Solid matrix compositions and spray drying process parameters are identified that enable RNA-loaded LNPs to be spray dried and reconstituted with minimal-to-no change in their critical quality attributes (CQAs) of particle size distribution, as measured by dynamic light scattering (DLS), and encapsulation efficiency of RNA, as measured by a RiboGreen fluorescence assay. We investigate the combinations and ratios of these components to best enable a target product profile defined for injectable delivery. The CQAs of spray dried and reconstituted LNPs from matrices composed of various combinations and ratios of these components are compared and implications to product design are discussed. Effects of spray drying process parameters, including liquid to gas mass ratio, atomization pressure, and outlet temperature, are also investigated. Special attention is paid to outlet temperature in order to avoid thermal stress and accelerated RNA degradation kinetics. A novel vacuum spray dryer design was employed to achieve outlet temperatures ranging from 5 – 60 °C. Finally, data is shown for optimized solid-state LNP formulations that result in minimal change to the particle size distribution of the LNPs upon spray drying and reconstitution, maintaining the CQAs of a Z-average diameter of less than 200 nm, polydispersity index less than 0.2, and RNA encapsulation efficiency greater than 80%. CQAs are tracked through accelerated stability studies to demonstrate the benefit of solid-state formulations for product storage and handling.

The results show that careful formulation of solid-state LNPs manufactured via spray drying can provide a scalable path to more widely available LNP-based nucleic acid therapies with improved shelf stability as a dry powder.

549c RNA Product Delivery Using 3D Printed Microfluidic Microarray Patches (MAPs)

Madison Driskill, Ian Coates, Jillian Perry, Shaomin M Tian, Yue Xu, Maria Dulay, Gunilla Jacobson, Netra Rajesh, Jonathan Smith, Joseph M. DeSimone

The use of 3D printing technologies, particularly single-digit micron resolution Continuous Liquid Interface Production (CLIP) developed by the DeSimone Lab at Stanford University, has promises to revolutionize microarray patch (MAP) production for drug and vaccine delivery. An additional significant 3D printing development for MAP production is the emergence of injection CLIP (iCLIP) technology, which enables the production of microfluidic MAP devices with double-digit micron features. These microfluidic structures can be integrated into the needles and backing of MAPs allowing for the creation of microfluidic structures within MAPs, leading to novel functionalities and enhanced capabilities. Additionally, this allows for rapid and precise delivery of drugs and vaccines. This advancement not only improves the efficacy of delivery, but also enables the administration of doses that were previously unattainable with conventional MAP platforms.

Moreover, the microfluidic backing of MAPs serves as a reservoir for lyophilized products, offering enhanced stability for shelf-unstable drugs and vaccines. This feature allows for the simultaneous reconstitution and delivery of these products, thereby ensuring their efficacy and viability during transportation and storage. We have demonstrated the utility of this approach for RNA delivery, achieving comparable reporter gene expression to intramuscular or intradermal controls.

The utilization of emergent RNA technologies, such as self-amplifying RNA (saRNA) encapsulated in lipid nanoparticles (LNPs), has further expanded the capabilities of the microfluidic MAP platform. By lyophilizing saRNA-LNP complexes, their thermostability is enhanced, enabling reliable and accessible vaccine delivery even in challenging environments. The combination of microfluidic MAPs with saRNA delivery holds promise for achieving equivalent immune responses with lower doses of RNA and longer duration of immunity, thereby potentially reducing manufacturing costs and improving vaccine accessibility. Additionally, being able to deliver lower doses of RNA to achieve equivalent immune responses also means that lower lipid doses are needed and lower reactogenicity may be achieved.

In summary, the integration of 3D printing technologies, microfluidic structures, and emergent RNA delivery methods has paved the way for the development of a robust, reliable, and thermostable vaccine delivery platform. This innovative approach has the potential to revolutionize vaccine distribution by offering improved stability, enhanced efficacy, and increased accessibility, particularly in resource-limited settings.

*This work was supported by the Wellcome Leap R3 Program and the Gates Foundation.

549d Rational and Combinatorial Design of Peptides for Ss-mRNA/dsRNA Separation and Purification

Mengyang Hu, Georges Belfort, Pankaj Karande

The development of a generalizable, fast, effective, scalable, translatable and economic purification platform is critical for the technical and economic success of mRNA-based therapeutics. Our goal in this work is to develop downstream bioseparation strategies to separate double-stranded RNA (dsRNA) impurities from single-stranded mRNA (ss-mRNA). To achieve this goal, we aim to identify peptide ligands that are selective towards dsRNA and couple these peptides to polymer filtration membranes to demonstrate the feasibility of purifying labile ss-mRNA from dsRNA at high yield and purity in a scalable fashion.

Two strategies were employed for the design and selection of peptide ligands. The first, a rational and focused approach, was based on mimicking the active binding pocket of naturally occurring dsRNA specific binding proteins. Protein design templates were selected based on the following criteria: (i) binding affinity to dsRNA; (ii) binding selectivity/specificity to dsRNA versus ss-mRNA; (iii) structural information available; and (iv) length/size of the protein. Four proteins/domains were identified as templates including a double stranded RNA binding domain and three RNA silencing suppressors. A library of 16-mer peptide candidates was generated using linear epitope mapping of the target proteins with particular focus on the peptides that cover the key interacting regions of the protein. A second approach based on combinatorial phage display was employed to screen a vast pool of randomized peptide sequences. Pure ss-mRNA and dsRNA were challenged against a commercial phage library for negative and positive selections. Lead peptide candidates identified from the two design strategies were synthesized and screened for dsRNA binding affinity and selectivity in an immobilized format on microarrays.

Peptides that exhibited strong binding to dsRNA while displaying weak or no binding to ssRNA were identified through high-throughput screening and further validated in an ELISA or microarray format. Lead peptides were mapped on to their parent protein templates and revealed a strong consensus with the dsRNA binding site. Interestingly, a few ssRNA-selective peptides were also identified, originating from the dsRNA-selective binding proteins but from different regions of the protein when compared to the dsRNA binding site. Lead sequences shared several common motifs and/or amino acid propensity.

Isolation of linker Intermediates for Antibody-drug Conjugates by Magnesium Complexation

Hochan Chang, Seung Wook Kim, Michael Dummeldinger, Nicolas Cuniere, Junhe Ma, Holly M. Shackman, Sharla Wood, and Shawn Brueggemeier*

E-mail: hochan.chang@bms.com (Presenting author, Corresponding author)*

Affiliation: Bristol Myers Squibb, Chemical Process Development, 1 Squibb Drive, New Brunswick, NJ, 08903

Extended Abstract

Due to properties of polyethylene glycols (PEGs), the modification of small molecules with PEG-chains can improve water solubility, stability, and oral bioavailability for pharmaceutical applications.[1] The synthetic route of a PEG4-containing linker, used in antibody-drug conjugates (ADCs), required a 7-step telescoped reaction with multiple chromatographic purifications because of the inability to crystallize or precipitate the PEG-containing molecules. Additionally, an intermediate of the linker had high viscosity, which hindered unit operations for material feeding and transferring upon process scale-up and prevented isolation of the stable intermediate in solid form to enable effective storage and transportation. To overcome these challenges, a new process was developed upon previous research, showing the formation of solid complexes from compounds that contain PEG-chains and acid functionalities by chelation of magnesium chloride ($MgCl_2$).[2, 3] However, the initial attempts to apply $MgCl_2$ complexation to this particular intermediate resulted in severe stickiness of the complex, which prevented practical application of the technique for process scale-up.

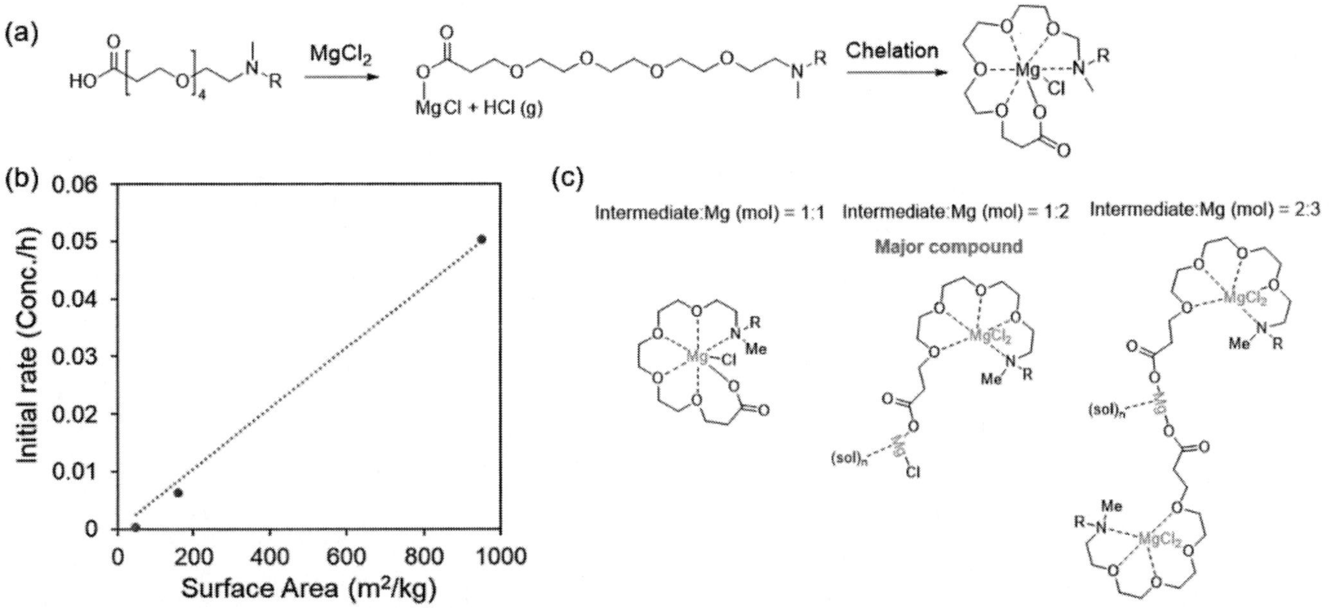

Figure 1. (a) Scheme of the complexation between $MgCl_2$ and PEG4-containing intermediate; **(b)** Effect of $MgCl_2$ surface area on initial rate of the complexation; **(c)** Structure of Mg-chelated compounds.

Herein, we utilized fundamental knowledge of chemical engineering to understand the complexation and to improve physical properties of the 'intermediate-$MgCl_2$ complex' (Fig.1a) for the development of a scalable process. Analysis of reaction kinetics revealed that the particle size of $MgCl_2$ (Fig.1b) and reaction temperature are the key parameters to complete the complexation without formation of the sticky crust, which inhibits recovery of the complex from the reactor. As a result, the use of milled $MgCl_2$ improved the physical properties of the solid complex and led a successful 100 L-scale process at 93 ± 2 % of recovery. Furthermore, various analytical technologies were used to elucidate the molecular structure of the complex. Elemental analysis, NMR DOSY experiments, and direct-infusion mass spectrometry supported the presence of complexes mixture with molar ratios of 1:1, 1:2 (majority), and 2:3 of the PEG4-containing intermediate to magnesium (Fig.1c). Accordingly, this isolated complex served as a stable intermediate with improved chemical purity and material handling, resulting in the enhanced scalability of the production process and the effective transportation of the material into the downstream processes. In downstream processes, pH-adjustment of the complex and extraction allow for the removal of $MgCl_2$ from the intermediate to continue the synthesis of the ADC linker.

References

(1) Turecek, P. L.; Bossard, M. J.; Schoetens, F.; Ivens, I. A. PEGylation of Biopharmaceuticals: A Review of Chemistry and Nonclinical Safety Information of Approved Drugs. *Journal of Pharmaceutical Sciences* **2016**, *105* (2), 460-475.

(2) Noto, V. D.; Münchow, V.; Vittadello, M.; Collet, J. C.; Lavina, S. Synthesis and characterization of lithium and magnesium complexes based on [EDTA][PEG400]2 and [EDTA]3[PEG400]7. *Macromolecular Chemistry and Physics* **2002**, *203* (9), 1211-1227.

(3) Zheng, B.; Li, J.; Pathirana, C.; Qiu, S.; Schmidt, M. A.; Eastgate, M. D. Complexation of Polyethyleneglycol Containing Small Molecules with Magnesium Chloride as a Purification and Isolation Strategy. *Organic Process Research & Development* **2021**, *25* (10), 2270-2276.

549f Isolation of Amorphous Peptides: Problems and Solutions

Paridhi Agrawal, Jennifer M. Groh, Ankur Jalan, Richard Miller

Natural and synthetic peptides, composed of a series of amino acids, are promising pharmaceutic modalities for the treatment of a wide variety of indications such as diabetes, cancer, migraine, heart failure and more. While producing peptides in a synthetic manner has several advantages such as being able to accommodate non-natural amino acids in the chain, it also comes with challenges around process complexity, elevated more so when following a hybrid approach that involves a couple dozen steps for API synthesis alone. To add to that, the drug substance manufacturing network is complex and spread around the globe, which often requires shipment of intermediates for downstream processing.

A common method for isolating solids that can be stored and shipped under stable conditions is freeze-drying or lyophilization. However, the lyophilization of peptides can sometimes be undesirable. Another way to isolate amorphous peptide intermediates and APIs is precipitation, which is easy to fit in any small molecule isolation equipment set.

Peptides are extremely prone to gelation. While the fundamentals of peptide gelation in organic media aren't completely understood to date, we have, in part, learnt to navigate the challenging physical properties and isolate peptides robustly.

In this presentation, we will discuss the various considerations in designing a peptide precipitation process such as the sensitivity around solvent selection, temperature, addition order and rate, ageing, and mixing. We will further share, through the example of a model peptide, the challenges encountered during precipitation, filtration and washing. We will present phase diagrams for process design and mixing models for scale up. Ultimately, manufacturing data showing the successful isolation of gelation-prone peptides will be shared.

281a Generation of Amorphous Solid Dispersions through Co-Precipitation for Enhanced Physical Properties

Tao Zhang, Corinne Jankovsky, Simone Blattner, Joe Gao, Xiao Cong, Shubhajit Paul, Huayu Li, Leon Schulz, Michael Braun

Due to the increasing number of new chemical entities that exhibit poor solubility, various strategies have been applied to improve the bioavailability, such as amorphous solid dispersions, salts/cocrystal formation, complexation techniques, lipid-based delivery systems, nanoscale drug delivery systems and particle size reduction. Converting the crystalline APIs into amorphous solid dispersions (ASDs) is gaining particular interest as it helps scientists address the challenges associated with poor drug solubility and ultimate leading to improved therapeutic outcomes.

ASDs often employ a polymer matrix to disperse the drug, which stabilize the amorphous state, preventing drug recrystallization. The amorphous state of the drug in ASDs can achieve higher apparent solubility, facilitate drug dissolution, and improve bioavailability. The common methods to prepare amorphous solid dispersions including spray drying and hot melt extrusions. The choice of the method depends on various factors including the drug and polymer properties, formulation requirements, and scale of production. Although ASDs offer numerous advantages, controlling physical properties of the resulting ASDs, such as the morphology, particle size distribution, density, and flowability, can be sometimes very challenging.

Herein, we developed a co-precipitation strategy to generate ASDs with well controlled morphology, uniform particle size distribution, high bulk density, and good flowability. The resulting materials were directly compressed into an oral tablet. The dissolution performance was compared against the spray dried dispersions (SDDs). *In vivo* results suggest the tablets can achieve equivalent pharmacokinetic performance.

281b Combined Impact of Process Parameters Temperature, Pressure, and Shear Stress on Polymorphism in Crystalline Solid Dispersion

Francheska M. Reyes Figueroa, José Hernández Espinell, Suresh Manivel, Lian Yu, Geoff G. Z. Zhang, Torsten Stelzer

Understanding the processing requireto generate crystalline solid dispersions (CSDs) containing metastable polymorphs
the application of hot-melt extrusion (HME) and 3D printing (3DP) as polymer-based formulation approaches. Hence, to successfully implement CSDs as alternative solid dosage formulations, the effect of critical process parameters (CPPs) on potential polymorphic phase transformations (PPTs) needs to be examined. This study extends the current knowledge on the influence of CPPs, through temperature-pressure-shear simulated extrusion (TPSS-E) using the model system flufenamic acid (FFA) in polyethylene glycol (PEG). The TPSS-E results revealed a significant reduction in the average PPT induction time (53 %) compared to previous temperature-pressure simulated extrusion (TPS-E) studies without shear stress. However, TPSS-E control experiments performed at 25°C did not show any PPT. uggestthat temperature is the most critical parameter on determining whether a PPT will occur, while pressure and shear stress significantly accelerates the PPT from the metastable () to the stable form (at elevated temperatures. These results demonstrate thatand it is possible to control the metastable polymorph in CSDs during extrusion processes if the thermodynamic and kinetic boundaries, the CPPs, and material attributes of the drug-polymer system are well understood and controlled. Ththe application of HME and 3DP as polymer-based formulation strategies for CSDs containing metastable polymorphs.

281c 3D Printed Dosage Development for Personalized Delivery of Nanodrugs to Pediatric Population

Nikki Rodriguez, Alperen Abaci, Gulenay Guner, Ecevit Bilgili, Murat Guvendiren

Children are not "little adults," which makes delivery of drugs to pediatric population challenging. Personalized delivery of drugs is especially pertinent as drug dose must be adjusted based on age and body weight of the pediatric patients. Besides ensuring efficacy, safety, and compliance, drug dosage must be pediatric-friendly, pain-free, and accurate. Unfortunately, traditional pharmaceutical solid and liquid dosages do not meet all these requirements. Up to now, tablets 3D-printed by fused deposition modeling feel like "hard plastics" and the production method is limited to a small class of thermally non-labile drugs. Finally, a great majority of the drug candidates in the pipeline of pharmaceutical companies are poorly soluble, whose delivery poses additional challenges. In this study, we present a robust formulation–process platform for direct ink writing (DIW) of soft nanodrug tablets that allow personalized delivery of poorly soluble drugs to pediatric population. Nanodrug suspensions prepared by wet media milling process are formulated into 3D printable inks. Ink rheology is correlated with ink flow and printability. 3D printed tablets with mechanical behavior comparable to commercially available gummies are selected, and further characterized for intra and inter content uniformity. Multi-material printing is used to create tablets with spatially controlled formulations. Multiple tablet designs are used to control API release kinetics. Overall, our robust and optimized 3D printing process–ink formulation can enable accelerated fabrication of personalized tablets composed of poorly soluble drugs.

281d Free-d Molding – Every Idea Deserves a Prototype

Daniel Treffer, Andreas Bramböck, Martin Genshofer

The ability to shape arbitrary geometries via 3D printing enables limitless opportunities to design drug-releasing products that are tailored to treat specific diseases at various positions inside human bodies [1]. The seeming beauty of 3D printing is that no complex manufacturing equipment is required; 3D printers are affordable as Fused Deposition Modeling (FDM) printers have become mass products over the last decades. Turning ideas into prototypes appears to be straightforward. However, for new 3D printed dosage forms, development goes beyond well-known feedstock material, facing challenges that seem almost impossible to overcome when approached together.

Major factors are (1) resources, (2) formulation development, and (3) process development.

The **resource challenge** gets tougher when working on new formulations. New Chemical Entities (NCE) can be expensive and unavailable to start a regular 3D printing project. The reason is simple. Conventional 3D printing usually requires closely defined material parameters, e.g., 1.75 mm filament diameter (+/- a few 1/100 mm) or a powder with defined particle size distribution and flowability. To obtain applicable starting materials, intensive upstream process development is a necessity. Thus, projects often end before they have a real chance to prove their success.

The **formulation challenge** is to choose the proper excipients from the many available options. Many vendors offer a wide variety of different polymers with different molar mass distributions or chemical modifications, each of which shows other material characteristics when it comes to processing, drug stability, and release; thus, many trials are necessary to identify the best carrier material. Running those trials with material-hungry extrusion equipment will waste a lot of material and valuable time as new formulations require time to find suitable processing parameters.

This is where the **process development challenge** starts. Small changes in API concentration can influence the material properties, causing the whole process to change its behavior. Finding the correct process parameters to achieve high product quality requires a lot of experience and material understanding, which is not available during the birth of an idea.

Efficient development is only possible when the complex development challenge is broken into two pieces: **formulation development** and **process development**. This reduces the complexity as interactions between both sets of challenges are minimized. The formulation challenge is the first to solve. Only if the formulation works, meeting target dissolution profiles, stability data, and other relevant product attributes, the processing challenges are worth solving. They are less complex as the formulation composition is well-defined and remains constant, and the process can be optimized for this specific case.

For the formulation development of simple geometries (e.g., discs, cylinders) Vacuum Compression Molding (VCM) has been established over the last few years to screen for amorphous solid dispersions [2] and drug-eluting implants. It is a powerful formulation screening tool for creating prototypes with little material knowledge. It mimics hot melt extrusion by the lossless melting of a homogeneous powder in an evacuated, adaptable sample chamber with minimal process parameters.

This case study demonstrates how the benefits of conventional 3D printing can be facilitated to generate molds for the VCM technology. The molds are inserted in the VCM chamber, and arbitrary shapes can be obtained with minimum process development as only processing temperature and time are necessary. This new method is called free-d molding, as researchers' ideas are freed [3].

METHODS

The first step starts on the computer by designing a 3D model of the desired dosage form. In the example below, a gummy bear is chosen. Then, straightforward 3D printing with engineering-grade resins creates positive models of the desired geometries. A non-stick material is used to cast negative molds from the 3D-printed parts, which can be placed inside the VCM Tool (MeltPrep, Austria). Since the whole VCM process operates under a vacuum, there is no air to displace inside the cavities of the molds. When heated, powdery materials fuse into melts, which fill the cavities. A detailed formulation prototype is created while running the conventional VCM process. Figure detail A shows a typical timeline for a free-d molding project from the idea to a new formulation prototype.

Depending on the shape and functionality of the final dosage form, three different types of free-d molding can be applied.

- For **imprint free-d molding**, a non-stick mold is inserted into the VCM Tool and filled with the powdery formulation. During the VCM process, the surface imprint on one side of the sample adapts to the used template, while the backside stays flat, forming a homogeneous disc. The amount of filled material adjusts the height of the backing.
- **Split-cavity free-d molding** uses two non-stick mold halves, forming one cavity with a gate facing towards the feed chamber holding the powdery formulation. During the VCM process, the formulation melts, flows through the gate, and fills the cavity to form any 3D geometry. The remaining material in the gate is called sprue and is removed with a clipper.
- **Multi-component free-d molding** is a multi-step process and can combine split-cavity and imprint free-d molded geometries. Various specimens from previous prototyping steps are fused to create one final formulation prototype.

RESULTS

To show the applicability of free-d molding, various matrix formulations and multi-component dosage forms used in the pharmaceutical field are tested. Feature sizes range from micrometers, enabling geometries such as microneedles with sharp tips, up to centimeters, allowing prototyping of floating devices or rings.

- **Imprint Free-d Molding** is used to demonstrate the prototyping of subcutaneous implants (Figure detail B, left). A batch with multiple rods was created in less than 15 minutes. Cutting off the implants with a flush cutter gave access to the individual drug-eluting implants with various cross-sections. Microneedle templates (Smicna, Singapore) enabled melt-based prototyping of microneedle arrays (Figure detail B, right).
- **Split-cavity molds allow complex 3D geometries.** Figure detail C shows a bone screw, rings, and various shapes for oral drug delivery. Depending on the geometry of the prototype, single pieces in round or elongated molds and multi-cavity molds are possible. The VitalDose example shows six rings of various sizes made in one chamber.
- **Multi-compartment free-d molding** was tested to replicate an example demonstrated by Triastek® (Figure detail d). The multi-compartment tablet contains different segments to load different formulations. Multicavity molds allow for making multiple pieces from one formulation per molding cycle. Subsequently, the individual components are assembled manually and fused via VCM to obtain the multi-component prototype.

CONCLUSION

3D printing is transitioning from an emerging technology to a robust manufacturing technique for next-generation pharmaceuticals with complex geometries. The novel free-d molding technique can enable challenging 3D printing projects right in the beginning. New geometries can be explored with minimum material knowledge and resources. This significantly increases the chances of success. Once the formulators meet the desired quality attributes, the project can move toward process development.

REFERENCES

1. Maniruzzaman, M. 3D and 4D Printing in Biomedical Applications, Weinheim, Germany: Wiley-VCH Verlag (2019)

2. Shadambikar, G. VCM as a Screening Tool to Investigate Carrier Suitability for HME, *Pharmaceutics*, (2020)
3. Repka, M. 3D Printing: Emerging Tech. and Functionality of Polymeric Excipients in Drug Product Dev, Springer (2024)

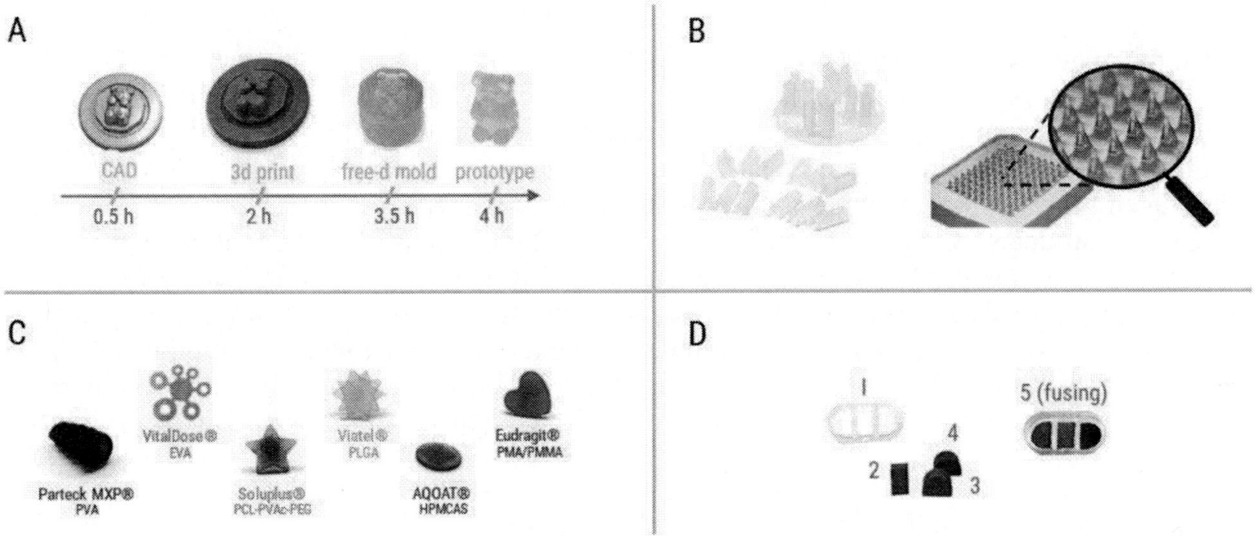

281e Assessment of Vacuum Drum Filling Technology Highly Cohesive Dry-Powder Inhaler (DPI) Formulation

Lucilla Coelho de Almeida, Kellie K. Sluga, Daniel Kestyus, Daniel Jakub Wilhelm, Justin Lacombe, Ariel Muliadi

Inhaled therapeutics are routinely used to treat pulmonary diseases such as asthma, COPD, and cystic fibrosis. In recent years, the need to deliver high therapeutic doses into the lungs is becoming increasingly more commonplace. In the case of recently approved anti-infectives (e.g. colobreathe, tobramycin DPI, aerovanc) and Parkinson's therapy (inbrija), for example, the total lung doses are thousands of times greater than those of conventional asthma medications (ICS, LABA, LAMA, etc.). Inhaled drug products, which are either a blend comprising the active pharmaceutical ingredient (API) and excipients, or engineered particles possessing attributes suitable for aerosolization (e.g. pulmospheres), are often packaged in blisters or standard-sized capsules and subsequently dispersed and delivered to the lungs by a specially designed inhaler. To promote patient compliance, it is desirable to administer the required lung dose in as few inhalations as possible. This, combined with the fill volume constraint of standard capsules, often leads to the need to maximize the drug loading in each dose.

This work presents the challenges associated with the process of filling a model high drug loading formulation into capsules. These challenges arise due to the cohesive and/or the low-density nature of the drug product, which is a consequence of the size and other attributes of the API payload: to reach the lung regions relevant to drug absorption, the API is typically micronized to a specific size range, or, in the case of engineered particles, their density is intentionally lowered by making the particles porous. While certain excipients can be blended with the API to improve filling capability, the fact remains that the amount that can be added is limited due to the reasons explained above. Such drug products, then, typically require specialized equipment to be filled into the dosage units.

Vacuum drum filling is one such equipment commonly used to manufacture dry powder inhalers (DPIs). Drum filling involves charging small amounts of formulation into a chamber wherein a drum into which small cylindrical cavities are drilled is placed on the bottom. When processing cohesive formulations, the charging of the formulation is often performed intermittently via a vibrating sieve placed on top of the chamber. Vacuum is then drawn from within the drum, causing some formulation to fill the cavities. The drum is then rotated by 180 degrees and air is blown out of cavities to eject the slightly compressed formulation out and into empty capsule bodies laid out under the drum. Throughout the drum dosing process, a stirrer placed inside the chamber is also intermittently rotating to prevent powder bridging with a rotational frequency that is in sync with the drum motion.

Despite its widespread use in this field, the drum filling process is not well studied. A literature search returns only two handfuls of articles, with most published in the last four years. To address such a lack, we performed a DoE study to investigate the effect of the process parameters hypothesized to affect the capsule filling process, namely the vacuum pressure, the blow-out pressure, the air pressure controlling the sieve vibration, and the rotational speed of the in-chamber stirrer. A ModuC-CS drum filler (Harro Hoefliger, Allmersbach, Germany) was used for the experiments. The following responses were measured: filling yield, fill weight consistency, capsule content uniformity (CU), and the aerosol performance of the filled capsules In addition, the amount of active ingredient in the formulation placed on the vibrating sieve is also tracked over time to assess potential vibration-induced segregations.

The results show high acceptance rates for all batches, ranging from 88-99%. The capability index, a measure of the capability of a process to provide output within the process specification limits, was between 0.56 and 0.99 for all runs, with the batch filled using a shorter filling duration and higher vacuum pressure/blowout pressure giving the highest dose consistency. Composite blend assays determined from samples taken from the vibrating sieve hopper at the beginning, middle, and end of each capsule filling run were within 97-103%, indicating no segregation risk for the formulation tested in this study. Similar trends were observed from the CU data assessed from capsule samples taken at the beginning, middle, and end of each run. All of the above results were also found to be relatively unimpacted by the in-chamber wire stirring speed, which directly affected the filling speed (i.e. faster stirring speed increases the drum rotation frequency, which in turn increases the overall filling speed). Finally, aerosol data assessed through the next generation impactor (NGI) and dosage unit sampling apparatus (DUSA) showed little to no influence of drum filling parameters on the aerosol statistics (e.g. emitted dose and fine particle fraction). All results confirm the feasibility and the robustness of a drum filling system for high-dose DPI manufacture.

281f Enhancing Material Characterizations for Tablet Development Using Split Instrumented Die Technology

Edward Yost

In the pursuit of advancing drug product manufacturing, this study employs a state-of-the-art die system facilitating triaxial decompression to investigate the critical unloading and ejection stages of pharmaceutical tablet production. Our experimental setup includes custom-built cylindrical dies integrated with piezoelectric sensors for precise measurements of die-wall pressures throughout the decompression process, and cuboid dies to assess the impact of shape on tablet properties. This combination of advanced instrumentation and die configurations provides unprecedented mechanistic insights into the decompression dynamics and their subsequent impact on tablet integrity.

Our examination spans commonly used excipients—microcrystalline cellulose, crystalline lactose monohydrate, and mannitol. The precise mapping of die-wall pressures facilitated by this technology effectively isolates the effects of unloading from other compaction variables. Although the triaxial decompression system can mitigate defects such as microcracks in certain formulations, its impact on mechanical properties, such as tensile strength, is minimal for materials that already produce defect-free compacts using a rigid die.

On the other hand, our data suggests that the split-able die-wall tooling allows better assessment of materials that are prone to defect formation when made using a rigid die. For these materials, dies with split-able walls enable the production of defect-free compacts, which are crucial for more accurately measuring true bonding strength of the materials; a quantity that better represents the materials' contribution to the bulk mechanical properties of the formulations that include these challenging materials. This enhanced level of material characterization, enabled by our novel instrumentation, offers valuable insights that could contribute to improved practices in material evaluation within the pharmaceutical industry, potentially giving way to better formulation design and optimization that enable faster 'first-time-right' tablet development for FIH studies.

281g Fundamental Motions: The Key to Accelerated Lyophilized Formulation Development

Kelly Badilla, Andreas Bommarius, Marcus T. Cicerone

Protein-based therapeutic drugs have served an important role in the treatment of several diseases and conditions, including influenza and certain cancers. Preservation of the proteins within such drugs is critical for patient safety and treatment efficacy. Lyophilization (freeze-drying) has been shown as an effective technique for enhancing the longevity of proteins, however, degradation is still imminent in many cases. While longer shelf-life is obviously advantageous, the result is long degradation studies to determine what formulations are most suitable for the stability of the protein within the drug product. In this work, I will demonstrate how probing "fundamental motions" in a lyophilized system can provide the insight required to accelerate lyophilized protein-based drug formulation development.

A 2012 publication by Cicerone[1] provided evidence for a relationship between the degradation rates of several proteins and the dynamics of the excipients that the proteins were lyophilized in. There are several works by others that investigate system dynamics through viscosity and glass-transition temperature, but to this end, show no clear correlation with stability. The publication by Cicerone, however, looked at more fundamental motions of the excipients: individual molecules hopping over each other. This data was acquired with neutron scattering, a technique less accessible to academia and industry, so the present work explores optical techniques to acquire similar data.

In this study, the long-term stability of lyophilized alcohol dehydrogenase (ADH), horseradish peroxidase (HRP), and IgG1 are characterized with enzymatic assays and size-exclusion chromatography. This data is then compared to the optical techniques which provide information regarding quantity of fundamental motions of the system (less motion means more stable protein). As these optical methods take on the order of minutes, with this work, we approach a promising method to reduce lyophilized protein formulation development times from a year to minutes.

1. Cicerone et al., Soft Matter, 8 (2012)

696a Numerical Analysis of Industrial Bioreactors to Accelerate Tech Transfer

Braden Chiulli, Homayon Homayoni, Maximilian Prachar

Biological medical products are gaining an increasing share of the pharmaceutical market, with 48% of newly approved drugs in 2023 being biologics [1]. This growth highlights the potential of biological medical products to treat patients with severe medical conditions. In this highly competitive landscape, there are mounting challenges to industrial biopharmaceutical companies, including: environmental issues, losses of revenue owing to patent expiration, increasingly cost-constrained healthcare systems, and more demanding regulatory requirements [2]. Given these industry headwinds, it is essential to optimize operations to increase speed to market while minimizing operating costs. Tech transfer and scale-up activities can contribute up to 20% of the workload for process development (PD) as time and resources are often allocated to executing demanding experiments [3]. To alleviate this resource demand and maintain sustainable, flexible development, a web-based application for accelerated tech transfer and scale-up in bioreactors is proposed.

The workflow consists of development of a Computational Fluid Dynamics (CFD) model for each bioreactor system, which is verified and used to complete a parametric study. The parametric study employs a design of experiments (DOE) approach to obtain data for the creation of mathematical models, enabling tech transfer and scale-up between bioreactors. This mathematical model is implemented in a graphical user interface (GUI), providing an easy-to-use application.

In the current initiative, five single-use bioreactors and two glass bioreactors are included: Cytiva Allegro STR 200, Cytiva Allegro STR 1000, Cytiva Xcellerex XDR 200, Cytiva Xcellerex XDR 2000, Sartorius Biostat STR 200, Sartorius Univessel Glass 2 L, Sartorius Univessel Glass 10 L. To accommodate the cellular oxygen demand, bioreactor operations must sustain a sufficient oxygen transfer rate (OTR). The OTR is dependent on the volumetric mass transfer coefficient $k_L a$, which can be calculated using Higgbie's penetration theory:

$$k_L a = C\sqrt{D_L}(\epsilon/\upsilon)^{0.25} a$$

Where D_L is the diffusion coefficient of a gas species in the liquid, ϵ is the turbulence dissipation rate, υ is kinematic viscosity, and is interfacial area concentration, which can be calculated as:

$$a=6\alpha/d_{32}$$

Where α is gas volume fraction and d_{32} is Sauter-mean diameter.

CFD models were developed to predict hydrodynamic behavior during cell cultivation in the mentioned bioreactors. The goal for the models was to reach a combination of required prediction accuracy and solution time. The CFD modeling was performed with Ansys Fluent R2021 software. The realizable k-ϵ model for turbulence and the Eulerian multiphase model were employed. The interfacial area was modeled using the interfacial area concentration approach with breakage and coalescence kernels. The rotation of the impeller was modeled using the multiple reference frame approach, and the case was solved as steady. The CFD model predictions were compared to the experimental data with mean relative error in the $k_L a$ values less than 15%. The P/V predictions have low variance; however, a bias was present due to different experimental setups, where the measurements were performed without gas sparging. The average solution time was approximately 1 day.

The CFD models were then used to perform a parametric study. The design of the experiment approach was selected, with three factors – working volume, agitation rate and gas flow rate studied at three levels (i.e., -50%, center-point, +50%). The center-point represents typical cultivation conditions used. The face-centered central composite design with 15 study points per bioreactor was selected to reduce the computational time without significantly diminishing information relative to the full-factorial DOE. The mathematical models for required hydrodynamic parameters were created using the locally estimated scatterplot smoothing technique.

The Shiny package in RStudio was used to develop an easy-to-use application. Scale-up and tech transfer are accomplished by fixing the oxygen transfer rate and a second parameter, which can be constant *vvm*, constant *P/V*, target agitation rate, target flow rate, hydrodynamic stress levels, or other additional terms. The application will be leveraged by process development and manufacturing experts to accelerate bioreactor scaling during tech transfer, reducing the need for resource-demanding wet-lab experiments.

Moreover, the CFD models developed in this work will be part of the "CFD library" for distribution within our R&D and manufacturing network. It is a library of all available CFD models for the equipment used, aiming to accelerate and improve the development of CFD models for specific purposes. Expanding this library to incorporate new vessels will enable more organizations and geographies to apply CFD to identify optimal operating parameters, perform facility fit evaluations, and even troubleshoot during runs.

References:

[1] M. Senior, Fresh from the biotech pipeline: record-breaking FDA approvals, Nature Biotechnology, 42 (2024) 355-361.

[2] S.M. Paul, D.S. Mytelka, C.T. Dunwiddie, C.C. Persinger, B.H. Munos, S.R. Lindborg, A.L. Schacht, How to improve R&D productivity: the pharmaceutical industry's grand challenge, Nature Reviews Drug Discovery, 9 (2010) 203-214.

[3] S.S. Farid, M. Baron, C. Stamatis, W. Nie, J. Coffman, Benchmarking biopharmaceutical process development and manufacturing cost contributions to R&D, mAbs, 12 (2020) 1754999

Figure 1 The operator inputs current process settings from the established process running on bioreactor A into the app. The app provides the operator with the input settings for the target vessel to maintain equivalent cultivation conditions while accepting user defined constraints.

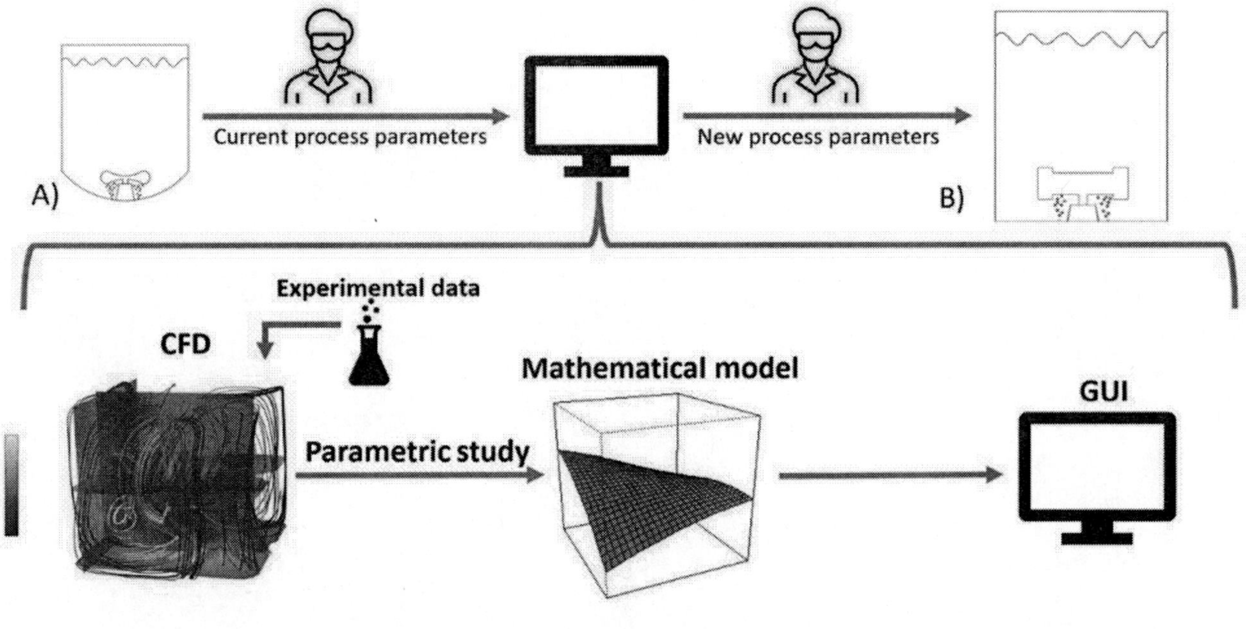

696c Enhancing Filtration and Chromatography Processes Using GPU-Accelerated Lattice-Boltzmann Simulations

John Thomas, Brian DeVincentis, Eric E. Janz

This presentation introduces a multi-scale approach to predicting two-phase flow through porous media, integrating lattice-Boltzmann (LB) simulations with fully resolved and semi-empirical Ergun models. By leveraging the LB method's suitability for complex topologies, the work provides insights into transport phenomena governing single and multiphase flow systems, with applications to filtration and chromatography processes. At small scales, this methodology involves creating packed beds using fully resolved discrete element method (DEM) simulations, enabling a broad exploration of flow behaviors crucial for optimizing filtration efficiency and chromatography separation processes. At larger scales, we consider Ergun models. We also address numerical considerations such as grid spacing optimization and simulation time step refinement to ensure the accuracy and reliability of predictions, facilitating comparison with experimental data. A key component of this analysis will focus on process scale-up, in terms of understanding physics across a range of operating volumes and scales.

One of the primary challenges in conducting high-resolution simulations lies in the computational burden associated with intricate models. To overcome this challenge and make our simulations industrially practical, we explore the utilization of graphics processing units (GPUs) for accelerating computations. GPUs offer parallel processing capabilities that significantly reduce computation time, enhancing the feasibility of our research and making it accessible for real-world industrial implementations.

696d Optimization and Scaleup of an Intensified Upstream Process

Zoe Horton, Jackie Gonzalez, Hideyuki Kajihara, Huimin Liao

Intensified fed-batch processes have the potential to both reduce production bioreactor duration and increase product titer. These processes are characterized by high seeding densities, typically greater than 5 million cells per milliliter. The high seed densities are achieved by implementing perfusion in the N-1 step. The use of perfusion and higher peak cell densities in the production bioreactor introduce operational and scaleup complexity through introduction of new variables and high oxygen requirements. We will highlight our work in optimizing N-1 and N unit operations to ensure the process is robust and scalable. We demonstrated the successful execution of this intensified upstream process with comparable performance across both bench scale and pilot scale operations.

696e Equipment Capability Curves for Large-Scale Freeze-Dryers Based on CFD Analysis

Matija Zaloznik, Blaz Kamenik, Matjaz Hribersek, Peter Dsouza, Johan Remmelgas, Johannes Khinast, Paritosh Pande, Mostafa Sulaiman, Efimia Metsi-Guckel, Avik Sarkar, Petr Kazarin, Frank Paolini, Francis DeMarco, Matej Zadravec

Equipment capability curves are important for designing freeze-dryers and are crucial for process design and scale-up. The capability curve describes the limit beyond which the geometry of the freeze dryer cannot support the flow of the sublimation stream from the drying chamber to the condenser. At the capability limit, the flow is thus choked by the geometry of the freeze dryer. The pressure in the drying chamber then increases and the sublimation rate decreases, which can lead to an increase in product temperature and a possible collapse of the porous cake to a dense plug.

In this study, equipment capability curves for three industrial-sized freeze dryers were derived from computational fluid dynamics (CFD) simulations. In the CFD model, the flow was modeled using a single-component, compressible, single-phase vapor in the laminar flow regime. Condensation on the condenser coils was models via additional desublimation sinks, and the influence of the condenser geometry on the capability curve was evaluated. The capability points defining the capability curve were calculated by setting a very low condenser pressure, a high temperature on the shelves in the chamber and different sublimate mass flows from shelves mimicking the ice slab test. The points of the capability curve are obtained by increasing the mass flow rate of the sublimate until the Mach number reaches 1 (point of choked flow). The pressure in the drying chamber is recorded and represents the result of the simulation, which defines the capability limit point at a certain mass flow rate of the sublimate.

Three lyophilizers were analyzed: two with side-mounted condensers and one with a bottom-integrated condenser. The simulation results are in good agreement with experimental results. This research thus contributes to the understanding, optimization and scale-up of freeze-drying processes using CFD and to the improvement of product quality and process efficiency in pharmaceutical applications.

696f Supporting Commercial Scale Lyophilization Technology Transfers Using Experiments and Modeling

Barton Dear

Typically, lyophilization cycles for pharmaceutical products are developed at lab scale due to many challenges associated with development at target scale (product limitations, costs, issues with timing, etc). However, because the heat and mass transfer properties that dictate the performance of the cycle are equipment-dependent (and particularly with the scale of the equipment), transferring a cycle developed at lab scale to commercial scale presents many new challenges that may detrimentally impact the product. This presentation explores how pairing select experiments with physics based and/or data-driven, statistical modeling can be used to support cycle transfer from lab scale to commercial scale ensuring robust, high quality product.

696g Integrating Mechanistic Modeling with Economic Analysis for End-to-End Bioprocess Design and Development

Behnam Partopour, Aditi Holey, Yuki Abe, Andrew Sinclair, David Pollard

The fast-growing field of biotechnology is at the cusp of a transformative leap where speed, economics, sustainability, and accessibility play essential roles. Therefore, early integration of mechanistic modeling with process/facility models is critical to evaluating and optimizing the economic and environmental impact of the manufactured product. This is essential for the holistic design and development of bioprocesses early in development, where there is the maximum potential to impact on cost and environmental outcomes.

This interdisciplinary approach marries the quantitative precision of mechanistic models, which elucidate the complex biological, chemical, and physical phenomena underlying bioprocesses, with the strategic foresight of economic analysis to assess cost, feasibility, sustainability, and risk. By embedding economic and sustainability considerations directly into the bioprocess design phase, this novel methodology enables the simultaneous optimization of both the technical, environmental, and financial aspects of therapeutic manufacturing. It facilitates the identification of cost-effective, scalable, and sustainable bioprocessing routes right from conception, thus significantly reducing development time, resources, and the risk of late-stage failures.

This presentation elucidates this integration's principles by showcasing its application through case studies that integrate mechanistic models for cell culture, harvest, and capture steps with cost and sustainability analysis within the BioSolve environment. Through multi-objective Bayesian optimization, we can identify the best solution space that minimizes cost and environmental impact, thereby providing insights into tradeoffs regarding cost and sustainability.

Integrating mechanistic and whole process models provides analysis that accelerates the path from lab to patients. It promises to enhance the viability and impact of biotechnological innovations in addressing global therapeutic challenges.

695a Application of Visible Particulate Controls for Long-Acting Nano Suspension Pharmaceutical Products

Christopher J. Morrison

All injectable products are required to be practically free of visible particulates. This is commonly achieved in solutions by passing the drug product through a 0.2 mm sterilizing grade filter just prior to filling. However, when a product consists of API particles in suspension that would immediately clog such a filter, a more holistic process approach is required to provide a level of assurance that the drug product suspension is practically free of visible particulates and microbial organisms while maintaining API content. To achieve this level of assurance everything entering the process must be considered. Areas of attention include API solution filtration prior to crystallization in drug substance, water activity in the API, 0.2 mm filtration of excipients once dissolved, API handling, API irradiation, closed processing, hold time, hold temperature, visible particulate filter with retention of 150 mm particles and above for suspension filtration, and terminal sterilization have been used successfully to deliver an industrial scale product.

695b Identifying Operating Conditions for the Solid-Phase Peptide Synthesis Process Via Optimization

Megan E. Walsh, Jennifer M. Groh, Mark Berglund, Jingyao Wang, Paridhi Agrawal, Casey Smith, Shekhar Viswanath, Chrysanthos Gounaris, Carl Laird

Synthetic peptides are an emerging product in the pharmaceutical industry being used to treat a range of illnesses due to their advantages over small and large molecule alternatives, including greater efficacy and lower toxicity[1]. The primary method used in industry to make synthetic peptides is the Solid-Phase Peptide Synthesis (SPPS) process which is a cyclic process where the desired peptide is built on a solid resin support one amino acid at a time. Each cycle of the SPPS process consists of a series of steps, including 1) Deprotection, where the Fmoc protecting group is removed to expose the amine on the resin or existing peptide chain on the resin, 2) Activation, where the incoming amino acid is activated, and 3) Coupling, where the activated amino acid couples to the deprotected end of the growing peptide chain on the resin.

The drawback to the SPPS method, however, is the generation of several impurities during the process, such as deletions, isomers, insertions, and capping adducts, that contribute to decreases in the purity and yield of the final product[1]. Given that many impurities are difficult to separate out in downstream processes, mitigating impurities during the build process becomes a main goal, but standard reagents and operating conditions to achieve this result in long build times and high process costs. Therefore, a decision-making framework is developed to provide users insight into the optimal operating conditions for a desired peptide build that will focus on impurity mitigation but also reduce long reaction times and high costs.

We have implemented an optimization-based decision-making framework in Python. In our previous work, we developed a detailed kinetic simulation in alignment with experimental results using the reaction mechanism for the SPPS process. This simulation was then numerically integrated to model the progression of all species in each cycle. The simulation allows the user to provide the operating conditions, such as the initial reagent charging, the reaction times for each cycle, and the intentional capping decision between cycles to help reduce impurity propagation throughout the build, for a desired peptide build and build metrics will be output, such as the peptide cost, throughput, purity, yield, etc. In this work, we tackle the inverse problem; given some objective, such as maximizing the throughput, minimizing the cost, or maximizing the purity of the desired build, the optimization framework solves for the optimal operating conditions to meet the objective while also adhering to strict purity requirements. Different methods of solving the optimization problem are explored, such as derivative-free optimization and surrogate-based approaches. Several benchmark cases are developed to show quantitative improvement over standard operation in reducing process time and costs while still ensuring the quality of the product[2]. Overall, this SPPS decision-making framework quantifies improvements in cost and process time reduction while still mitigating impurity generation all while providing users with a quick way to determine operating conditions that can be further explored experimentally.

(1) Frederick, M. O.; Boyse, R. A.; Braden, T. M.; Calvin, J. R.; Campbell, B. M.; Changi, S. M.; Coffin, S. R.; Condon, C.; Gowran, O.; Groh, J. M.; Groskreutz, S. R.; Harms, Z. D.; Humenik, A. A.; Kallman, N. J.; Klitzing, N. D.; Kopach, M. E.; Kretsinger, J. K.; Lambertus, G. R.; Lampert, J. T.; Maguire, L. M.; Moynihan, H. A.; Mullane, N. S.; Murphy, J. D.; O'Mahony, M. E.; Richey, R. N.; Seibert, K. D.; Spencer, R. D.; Strege, M. A.; Tandogan, N.; Torres, F. L. T.; Tsukanov, S. V.; Xia, H.

Kilogram-Scale GMP Manufacture of Tirzepatide Using a Hybrid SPPS/LPPS Approach with Continuous Manufacturing. *Org. Process Res. Dev.* **2021**, *25* (7), 1628–1636. https://doi.org/10.1021/acs.oprd.1c00108.

(2) Walsh, M. E.; McClary Groh, J.; Berglund, M.; Wang, J.; Agrawal, P.; Smith, C.; Viswanath, S.; Laird, C. D.; Gounaris, C. E. Simulation-Based Optimization of the Solid-Phase Peptide Synthesis Process. **2024**, *Forthcoming.*

695d Developing an Automated Bioreactor Feedback Control Toolbox for Cell Culture Processes

Chris Lowe, Gioia Tuozzolo, Benjamin Bayer, Leo Karmiy, Zach Houle

Process analytic technology (PAT) has long held promise to deepen understanding of cell culture processes. Great advancements have been made in extracting additional information through online probes and at-line analytical equipment but in current deployments, that data is often under-utilized. Cell culture processes have long utilized in-process criteria to trigger process actions but still rely on operators to interpret and manually input process changes such as feed delivery, temperature shifts, gassing adjustments, and others. To fully leverage the increased process information provided by PAT technologies, feedback control should be established between at line analyzers or online probes and the bioreactor control systems.

This presentation will describe automated tools developed by Takeda's Biotherapeutics Process Development team, both with existing biologics infrastructure and emerging PAT tools, and discuss their potential for deployment across development labs, pilot plants, and GMP manufacturing spaces. Closed loop bioreactor feedback control enabled by integration of bioreactor control software (DeltaV), auto sampling infrastructure (MAST or Nova Flex2 OLS), and cell culture analyzers (Nova Flex2) already present in process development laboratories has reduced hands on operator time by >90% for routine sampling and feeding activities. These tools are implementable with existing instruments and infrastructure and have been successfully deployed across 38 bench scale bioreactors in the process development labs and to large scale bioreactors (50L, 200L, and 500L) in the pilot plant. In addition to automation of daily sampling and feeding operations, manual process intervention for temperature shifts, feed initiation, or perfusion rate modifications has been automated using online capacitance probes. A soft-sensor model was developed to translate the online capacitance signal to offline cell counter measurements on the Flex2. This model was implemented into the DeltaV bioreactor control software (1) to directly control perfusion rate in an n-1 perfusion bioreactor and (2) to automatically initiate temperature shift and feed delivery in a fed batch bioreactor process. These advances allow for increased process consistency, by initiating process changes automatically when the culture dictates, regardless of operational constraints, and allows development labs to approximate 24-hour coverage without the need for 24-hour staff. Additionally, the combined implementation of these tools allows for end-to-end hands-off operation of a fed-batch bioreactor where automated features maintain the bioreactor entirely from inoculation to harvest. Collectively this set of automated tools not only allows for reduced operational burden but also greatly reduces the possibility of human error during bioreactor sampling or in interpretation of in process data and action criteria.

Continued development of these automated tools is enabling flexible, fit for purpose deployment across development and manufacturing spaces. While throughput gains and reduced operational resources are needed in the small-scale process development labs, reliability, process consistency, and reduced operational risk are critical for manufacturing spaces. These automation initiatives not only provide immediate benefit to the end users, but also better prepares these spaces for future adoption of advanced process tools as they continue to evolve. Overall, this suite of automated tools allows for flexible implementation to meet the needs of different facilities and provides a foundational backbone for advanced processes that will be enabled by PAT in the future.

695e End to End Cell Culture Modeling: Improving Process Development Agility. Starting and Ending With Data

Matthew Demers, Miguel Angel Valderrama-Gomez, Taehun Kim, William Heymann III, Juliane Glaser, Sidra Xu, Elcin Icten Gencer

Cell culture processes are widely used in biotechnology and pharmaceutical industries to produce recombinant proteins, monoclonal antibodies, vaccines, and other biologics. These processes are complex and nonlinear, involving multiple interacting factors that affect cell growth, metabolism, and product formation. Designing and optimizing these processes to achieve high productivity and meet quality criteria is time and resource intensive. *In silico* modeling can be used in the process development workflow to speed up development time and reduce the need for experimentation. Kinetic modeling is one approach that has been successfully used to simulate these complex processes.

In this work, a kinetic model capable of simulating and optimizing fed batch and perfusion-based cell culture processes is presented. This model is derived from the work by Kontoravdi et al. [1] and focuses on the metabolic changes and dynamic character of cell culture processes, especially with complex feeding strategies. Accurately representing lactate metabolism is a central step in model design due to lactate's important role in overall cell metabolism as an energy source and a growth inhibitor. To properly account for lactate as a viable energy source, its consumption must be accounted for in the model's kinetic and mass balance layers. The consumption of lactate is dependent on many factors, such as glucose concentration, which can make it difficult to accurately model using conventional kinetic methods. Here, we represent lactate's consumption and its central role in cell growth by using Michaelis Menten and more complex rate laws which is crucial in processes with complex feeding strategies due to resulting fluctuations in lactate production and consumption. The model performance will be compared versus Kontoravdi et al.'s model using a variety of analyses to understand model accuracy, robustness, and capability. This lactate implementation provides additional value to the *in silico* process development workflow by enabling high simulation accuracy and adding further mechanistic principles.

References:

[1] Kontoravdi C, Pistikopoulos EN, Mantalaris A. Systematic development of predictive mathematical models for animal cell cultures. Computers & Chemical Engineering. 2010;34(8):1192–1198. doi:https://doi.org/10.1016/j.compchemeng.2010.03.012.

695f Enhancing Sterile Filtration of Nanoemulsions Using Prefiltration

Shreya Kapila, Randy Soukup, Marissa Bradley, Andrew Zydney

Nanoemulsions are currently used as adjuvants in the development of novel vaccines and as an advanced method of drug delivery which act as a delivery system for hydrophobic drug products. Nanoemulsions are prepared by a multi-step process involving the mixing of immiscible aqueous and organic phases, typically in the presence of one or more surfactants, resulting in a suspension with nano-sized droplets with diameters of 20 to 200 nm. This can create significant challenges for the final sterile filtration step which uses sterilizing grade filters that are rated with a 200 nm pore size. The objective of this study was to examine the sterile filtration of a model nanoemulsion, with a specific focus on the opportunity of enhancing the product yield and filter capacity by appropriate choice of a prefilter.

Experiments were performed with nanoemulsions made using squalene with Tween 20 as a surfactant. Sterile filtration was performed at a constant filtrate flux, with the transmembrane pressure monitored as a function of the volumetric throughput. A number of sterile filters and prefilters were examined with different pore morphologies and surface chemistries. In each case, a significant transmembrane pressure (TMP) was required to obtain any flux through the membrane, with typical values for the Pall Supor (polyethersulfone) dual-layer filter of around 25 psi. This large initial TMP reflects the force required to deform and "push" the nanoemulsion droplets through the small pores of the sterile filter. The initial TMP could be significantly reduced by prefiltering the nanoemulsion through an appropriate pore size prefilter, with the best performance obtained using 0.45 μm polyvinylidene fluoride (PVDF) prefilters. The TMP increased nearly linearly with volumetric throughput, with the slope, and thus the capacity, dependent on the properties of both the sterile filter and the prefilter. Contact angle measurements using the sessile drop method with water revealed a contact angle of 120° ± 3° ($\theta > 90°$) for Sterlitech PVDF, indicating it possesses a hydrophobic surface. Even though the water permeability of sterlitech PVDF prefilter is very less (< 300 LMH/psi) compared to > 2000 LMH/psi for PES filters owing to their less hydrophilic surface properties. Bubble point tests reveal it has the largest pore size (1.72 μm) despite its rating as 0.45 μm, compared to all other prefilters. Most hydrophobic prefilters, like PVDF, showed lower initial TMP and higher capacity, possibly due to their less hydrophilic nature, as indicated by contact angle measurements and XPS analysis, aiding in easier passage for oil droplets through the filter. Thus, a combination of differences in pore size and surface properties play a role in determining the prefiltration performance. These results provide important insights into factors controlling the sterile filtration of nanoemulsions as well as a framework for the design and optimization of improving sterile filtration processes.

695g Facing Post Process Validation Challenges in Drug Product Manufacturing: One Biological Product, Two Case Studies

Sheetal Pai-Wechsung

Drug product process validation is a key milestone in the successful development of biological therapeutics. Unforeseen post process validation challenges could significantly delay the time to market, and the ability to devise a timely and robust strategy to identify the root cause and mitigate the risks is crucial to course correct. The presentation will highlight the practical approaches that led to overcoming technical and strategic challenges through two case studies for a multi-dose biological drug product. The first case revolves around the loss of preservative at the end of the proposed shelf-life. The second case explores an increase in high molecular mass species in drug product batches containing scaled-up drug substance. The multi-prong mitigation approaches to support post process validation manufacturing, including the use of the Six Sigma DMAIC data-driven improvement cycle, will be discussed.

695h Nanomilling with CPS Beads for the Production of Drug Nanosuspensions: Heat Generation and Particle Breakage

Hamidreza Heidari, Patrick McAleer, Gulenay Guner, Don Clancy, Ecevit Bilgili

Wet stirred media milling (WSMM) has emerged as an indispensable unit operation in the pharmaceutical industry to produce drug nanosuspensions, offering numerous advantages such as increased dissolution rates, enhanced bioavailability, and improved physical and chemical stability of poorly water-soluble drugs. However, the WSMM process is associated with a significant challenge: the generation of substantial heat due to dissipation of high mechanical energy input. Excessive heat generation and associated high temperatures during milling can lead to adverse consequences, including thermal degradation of drugs, solid-state changes of drugs, and destabilization of nanosuspensions, ultimately compromising the quality and stability of the final drug product [1]. Only a few studies have aimed to maintain the milling temperature below a set value, often employing intermittent milling strategies to mitigate excessive temperature rise during WSMM [2]. While previous studies have indicated the issue of high heat generation during WSMM, a comprehensive investigation of the impact of process parameters on temperature rise has attracted little attention [1]. Guner et al. conducted the first investigation into the influence of process parameters on heat generation when zirconia beads were used in WSMM [2].

In this study, we undertook a comprehensive investigation of the effects of three critical process parameters – stirrer speed, bead loading, and bead size – on the evolution of mill outlet temperature during the milling of fenofibrate (FNB) suspensions using crosslinked polystyrene (CPS) beads. The choice of CPS beads, with their lower density compared to the traditionally used zirconia (YSZ) beads, was motivated by the hypothesis that the use of lower-density beads could potentially mitigate heat dissipation and facilitate better temperature control during the milling process. Particle sizes, temperature evolution, and net power consumption were measured. The aqueous suspension formulation was fixed: 10% FNB, 8% hydroxypropylcellulose (HPC, L-grade), and 0.05% sodium dodecyl sulfate (SDS) in deionized water. Here, FNB is a poorly soluble, model drug; HPC and SDS serve as stabilizer and wetting agent. A microhydrodynamic model was used to elucidate the impact of the process parameters. The temperature at the mill outlet was aimed to be below 45°C to avoid potential gelation of the stabilizer (HPC-L) [1].

Salient features of temperature evolution during the milling experiments were as follows: a monotonic increase in temperature with a decreasing rate was observed with the attainment of a steady-state outlet temperature. This can be explained by heat generated by various dissipative processes, including bead-to-bead and bead-to-wall collisions, as well as viscous heat generation [3,4]. Higher power consumption, facilitated by higher stirrer speeds, higher bead loadings, and larger beads, resulted in faster temperature rise and higher maximum temperatures. However, a notable finding was that even at the highest power consumption, the maximum temperature rise was limited to 25°C, eliminating the need for intermittent milling, which is a common practice when using zirconia (YSZ) beads under medium-to-high energetic conditions.

Power-law correlations were established, relating the normalized temperature rise and the power consumption to the process parameters. The analysis revealed that the stirrer speed had the most significant impact on power consumption and temperature rise, followed by bead loading and bead size. The established power-law correlations provide a quantitative understanding of the heat generation–temperature rise dynamics during WSMM, enabling the prediction and optimization of process parameters to maintain desired temperature ranges.

To gain deeper insights into the intricate relationships between process parameters, particle size reduction, and heat generation, we employed the microhydrodynamic (MHD) model [3,4]. It is based on the kinetic theory of granular flows and fundamental granular energy balance [5]. Various MHD parameters, such as granular temperature, average bead oscillation velocity, and average frequency of drug particle compression were calculated at different processing conditions. It was found that bead loading had the highest impact on most MHD parameters, followed by the stirrer speed, while the bead size had a relatively weak effect. A correlation was established between the median particle size and the average frequency of drug particle compression, indicating a strong relationship between these two variables. Furthermore, a power-law correlation was developed between the dimensionless temperature rise and the MHD parameters, providing a quantitative understanding of the interplay between heat generation, particle size reduction, and process parameters. The microhydrodynamic analysis further elucidated the intricate relationships between process parameters, particle size reduction, and heat generation, highlighting the pivotal role of the average frequency of drug particle compression in governing particle breakage kinetics [6].

In a notable comparative analysis, we juxtaposed the performance of CPS beads with previously reported results [1] with zirconia (YSZ) beads under similar conditions. **Figure 1** illustrates a comparative analysis of the temperature rise of the suspension at the mill outlet when utilizing YSZ beads and CPS beads, considering various bead loadings. The stirrer speed was maintained at 4000 rpm, and the bead size was set at 400 m. The findings revealed that the use of CPS beads led to lower heat generation rates and lower temperature rise during the milling process, while still producing fine, stable fenofibrate suspensions with comparable particle size distributions. Remarkably, the maximum temperature attained during the process was consistently lower when using CPS beads compared to YSZ beads. This temperature difference became more pronounced at higher stirrer speeds and bead loadings, further highlighting the distinct heat dissipation characteristics of the two bead types. Most remarkably, the use of CPS beads eliminated the need for intermittent milling, even at higher stirrer speeds, which was a necessity when using YSZ beads due to excessive temperature rise in the lab-scale mill [7,8]. For YSZ beads, Figure 1 illustrates a multitude of cycles; each cycle consists of milling–cooling up to 45 ^{o}C followed by cooling to ~18 ^{o}C. There was only one cycle for the CPS beads. These findings underscore the advantageous use of CPS beads for low heat generation and easy temperature control during the wet media milling of thermally labile drugs, where strict temperature control is critical for maintaining product quality and stability.

In conclusion, we have demonstrated that the conversion of mechanical power into heat led to temperature rise during the WSMM process, and stirrer speed was the most influential parameter affecting temperature rise, followed by bead loading and bead size. Moreover, this study contributes to the understanding of heat generation dynamics in WSMM and provides valuable insights into process optimization strategies for temperature control. We have demonstrated that CPS beads may be preferred over traditional YSZ beads to produce nanosuspensions of temperature-sensitive drugs, offering potential benefits in terms of product quality, process efficiency, and energy conservation.

References

[1] Guner, G., et al., Analysis of heat generation during the production of drug nanosuspensions in a wet stirred media mill. International Journal of Pharmaceutics, 2022. 624: p. 122020.

[2] Afolabi, A., O. Akinlabi, and E. Bilgili, Impact of process parameters on the breakage kinetics of poorly water-soluble drugs during wet stirred media milling: A microhydrodynamic view. European journal of pharmaceutical sciences, 2014. 51: p. 75-86.

[3] Bilgili, E. and A. Afolabi, A combined microhydrodynamics–polymer adsorption analysis for elucidation of the roles of stabilizers in wet stirred media milling. International journal of pharmaceutics, 2012. 439(1-2): p. 193-206.

[4] Eskin, D., et al., Microhydrodynamics of stirred media milling. Powder Technology, 2005. 156(2-3): p. 95-102

[5] Gidaspow, D., Multiphase flow and fluidization: continuum and kinetic theory descriptions. 1994: Academic press.

[6] Guner, G., et al., Effects of bead packing limit concentration on microhydrodynamics-based prediction of breakage kinetics in wet stirred media milling. Powder Technology, 2022. 403: p. 117433.

[7] Guner, G., et al., Use of bead mixtures as a novel process optimization approach to nanomilling of drug suspensions. Pharmaceutical Research, 2021. 38(7): p. 1279-1296.

[8] Parker, N., M. Rahman, and E. Bilgili, Impact of media material and process parameters on breakage kinetics–energy consumption during wet media milling of drugs. European Journal of Pharmaceutics and Biopharmaceutics, 2020. 153: p. 52-67.

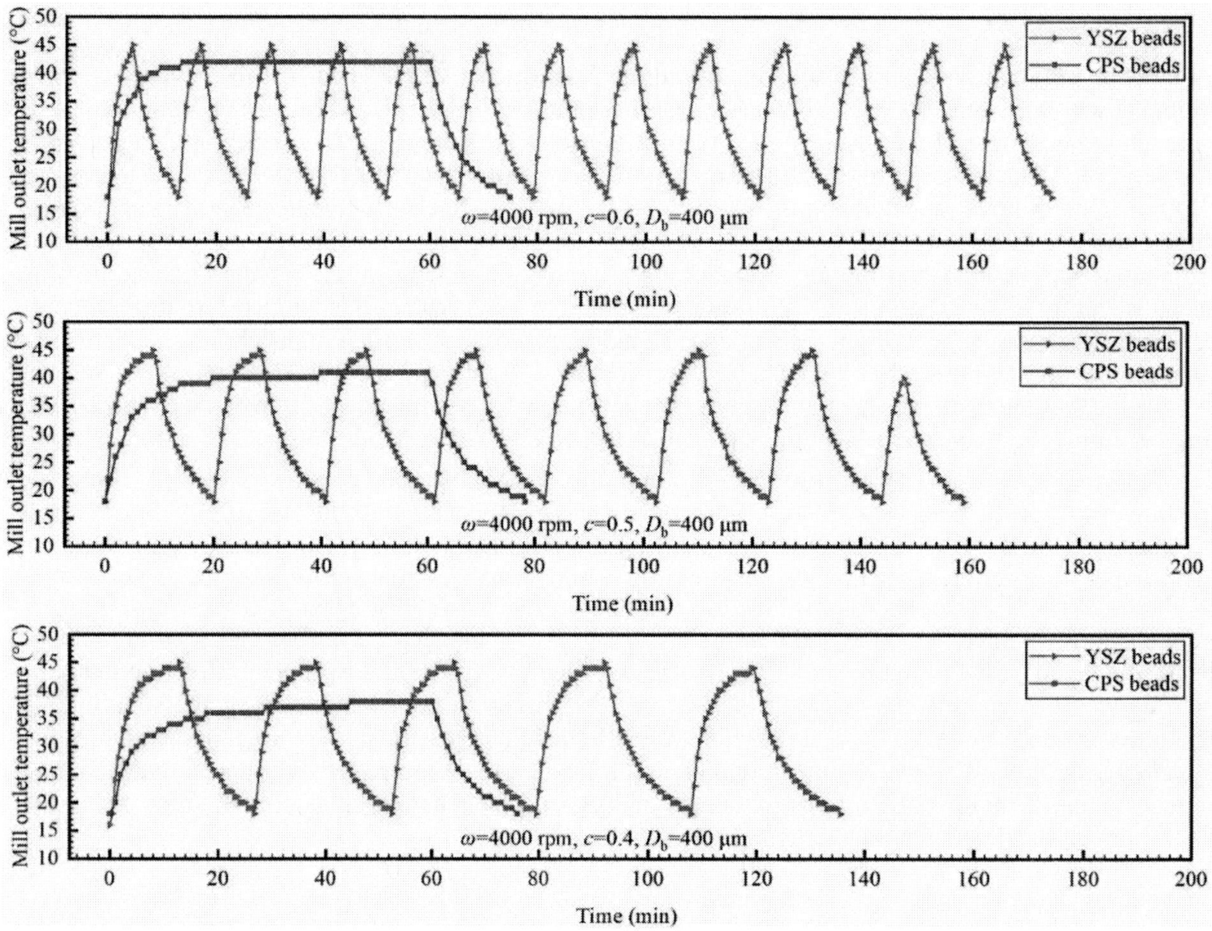

Figure 1 – A comparison of temperature rise in suspension at the mill outlet using YSZ beads and CPS beads at various bead loadings, with a stirrer speed of 4000 rpm and a bead size of 400 μm.

655a Defining Suitable Process Operating Range for a Spray Dried Dispersion Using Combined Mechanistic and Statistical Modelling Tools

Suela Jonuzaj, Christopher L. Burcham, Erin D. Copeland Jones, Kerri Fischer, Paulo Francisco, Lexie Niemoeller, Chad Wolfe

Spray drying dispersion (SDD) is a commonly used technology in drug manufacturing to improve solubility and dissolution behaviour of poorly soluble compounds. Designing a robust spray drying process and defining a feasible operating space at commercial scale are important in the highly regulated pharmaceutical industry. The use of computational tools can facilitate process development and design space analysis, while reducing experimental investigations and material consumption. In this work, we present a model-based workflow for spray drying processes, where a mechanistic model is utilized to identify key process parameters, such as gas/solution flowrates and processing temperatures, at large scale. Next, we evaluate a wide range of operating conditions and their impact on the physical stability of the SDD using data-driven correlations integrated within the mechanistic model, which allowed for the prediction of a broad operating space. Finally, a statistically designed experiment was used to determine the relationship between key process parameters and physical property attributes. The data from a reduced and well-defined set of experiments confirmed that desired quality attributes were achieved with a wide operating space.

655b Establishing a Powder Characterization Workflow to Determine Ideal Process Parameters for a Vibratory Mixing Process

Andreas Kottlan, Benjamin Glasser, Johannes Khinast

Blending of granular media can be deemed as one of the fundamental unit operations in the tablet manufacturing. The homogeneity of a powder blend does not only influence content uniformity in most process schemes, but also the proper function of all excipients being part of a drug product. The present study considers a process route where this issue is circumvented by design via dosing the ingredients for each tablet one after another into the tableting die prior to blending and compaction. Nevertheless, the homogeneity of the blend is crucial for obtaining a tablet exhibiting desired critical quality attributes (CQAs) like hardness or release characteristics. The abovementioned process route, i.e., single-tablet-scale direct compression[1] relies on a vibratory mixing process for blending small amounts (< 1g) of powder. While this process is fast and highly efficient for free-flowing to moderately cohesive powder systems, the process parameters for blending cohesive mixtures need to be picked with care. This study proposes a systematic approach to derive the ideal vibration parameters for blending via standardized shear cell experiments. The prediction workflow relies on a semi-mechanistic model which was derived based on high-speed imaging investigations of the powder bed in the mixing vessel which revealed the failure mechanism of the vibratory mixing process. High-speed imaging revealed that the high accelerations inherent to a fast vibratory mixing process lead to the formation of a densely packed layer, if the mixture contains a relevant amount of cohesive powder. If the strength of this layer exceeds the stresses imposed on it by the vibratory mixing process, it cannot be disrupted, which prevents the agitation of the powder bed. The semi-mechanistic model developed in this study predicts the strength of this powder layer based on the cohesivity of the ingredients, determined in shear cell experiments, and the compaction pressure imposed on the powder bed during the oscillatory motion of the mixing vessel. A similar approach which is used for determination of the compaction pressure is used to calculate the tensile stress emerging in the powder bed during oscillation, enabling the calculation of the operating window for this powder system. Together with information from previous studies of Kottlan et al.[2,3] , which suggest, that the within a range of 100-300 Hz, lower frequencies lead to faster mixing, the ideal frequency can be determined. Experimental validation with the cohesive excipients Sorbolac 400 and Vivapur 105 showed good correlation with the prediction. The formation of a stable layer was not observed at the frequencies suggested by the model, i.e., 120 Hz and 130 Hz respectively, enabling the mixing process for these cohesive excipients.

Hence, this study lays the basis for predicting reliable process parameters for small-scale vibratory mixing, while providing unprecedented insight to the complex behavior of cohesive powder in this very process.

References

1. Kottlan, A. *et al.* Single-tablet-scale direct-compression: An on-demand manufacturing route for personalized tablets. *Int J Pharm* **643**, (2023).
2. Kottlan, A., Glasser, B. J. & Khinast, J. G. Vibratory mixing of pharmaceutical powders on a single-tablet-scale. *Powder Technol* **387**, 385–395 (2021).
3. Kottlan, A., Glasser, B. J. & Khinast, J. G. Powder bed dynamics of a single-tablet-scale vibratory mixing process. *Powder Technol* **414**, 118029 (2023).

655c Development of a Solubility Measurement Workflow for Accelerating Crystallization Process Development for Pharmaceutical Compounds

Varun Sundarkumar, Laura Pereira Diaz, Yan-Shu Huang, Adrian Ramirez Galilea, Charles D. Papageorgiou, Neda Nazemifard

Crystallization is the primary method for isolation of small-molecule active pharmaceutical ingredients (APIs) and intermediates. Typically, separation of API from a solvent mixture is achieved by either lowering solution temperature, dosing an antisolvent – a substance with low solvation for the active, adjusting pH, or a combination of these methods. A robust and well-developed crystallization process has several characteristics, such as delivering product with high purity, favorable yield, and forming stable polymorphic structures. Additional qualities include having a low process mass intensity (PMI) ratio, using solvents with low toxic potential, economic efficiency, and ease of control. Consequently, developing a robust crystallization process presents a complex problem demanding significant time and resources (Brown et al., 2018).

A key aspect of developing a crystallization process involves the careful selection of solvents and antisolvents. Good solvents exhibit higher solvation capacity for process impurities than the active ingredient, thereby facilitating the purging of impurities and isolation of pure crystals. They also show a steep change in API solubility with temperature and thus maximize crystal yield with the lowest possible PMI. Similarly, effective antisolvents must mix well with the solvent and have minimal solubility for the API. Traditionally, solvent selection is conducted through a screening process, by manually testing several candidate solvents and antisolvents based on experience. This approach is labor, time, and material intensive and may yield sub-optimal solutions. Developing an efficient workflow for solvent screening and solubility measurement can thus accelerate crystallization development and lead to optimal processes with the above-mentioned characteristics. Several studies, in academia and industry, have explored the development of such workflows. These encompass the development of model-based tools to predict solubility ab-initio (Klajmon, 2022; Nordström & Rasmuson, 2009), automated high throughput platforms that can rapidly test several combinations of solvents (Shiri et al., 2021), and hybrid approaches combining these methods (Tan et al., 2019).

This study presents the development of a solvent screening and solubility measurement workflow for Takeda to support crystallization development for pipeline projects. This workflow is designed to minimize material usage, accelerate development timelines, and enhance overall process design. It will be hosted on an intra-group web platform and tailored to be user-friendly for chemists and engineers of all experience levels. The workflow will integrate model-based predictions, high throughput screening, and detailed experimental measurements for solubility (figure 1). At each step, multiple model packages and screening solutions will be used to ensure the workflow's adaptability across different application areas and to facilitate the integration of latest technologies.

The first step in the workflow is to generate a list of feasible solvents and antisolvent for the API by performing a structure-based in-silico prediction of solubility. In the next step, API solubility is measured in these feasible solvents at room temperature. This data is used to refine the prediction made in the previous stage and generate a candidate solubility curve. Potential crystallization conditions are determined based on this solubility curve, such as solvent-antisolvent pairs, temperature ranges, expected yields and impurity purging. The scientist can then select the most suitable option among these and proceed with detailed solubility measurements. At this stage, experimental and predicted solubility can be compared, and further exploration can be carried out in case of substantial deviations. Finally, a detailed solubility model, specific to the system, is built through regression. This model can be used to determine crystallization conditions such as seeding point and can serve as a fast-executing model for process control.

The proposed workflow will be validated for three test APIs – two proprietary compounds and Paracetamol. The development of this workflow will thus create a systematic guide for solvent selection and solubility measurement and accelerate the development crystallization operations. Additionally, by taking into account factors impurity rejection, solvent toxicity, PMI, and product stability, it will facilitate the development of more optimal processes.

References

Brown, C. J., Mcglone, T., Yerdelen, S., Srirambhatla, V., Mabbott, F., Gurung, R., Briuglia, M. L., Ahmed, B., Polyzois, H., Mcginty, J., Perciballi, F., Fysikopoulos, D., Macfhionnghaile, P., Siddique, H., Raval, V., Harrington, T. S., Vassileiou, A. D., Robertson, M., Prasad, E., ... Florence, A. J. (2018). Enabling precision manufacturing of active pharmaceutical ingredients: workflow for seeded cooling continuous crystallisations. *Molecular Systems Design and Engineering*, *3*(3). https://doi.org/10.1039/c7me00096k

Klajmon, M. (2022). Purely Predicting the Pharmaceutical Solubility: What to Expect from PC-SAFT and COSMO-RS? *Molecular Pharmaceutics*, *19*(11). https://doi.org/10.1021/acs.molpharmaceut.2c00573

Nordström, F. L., & Rasmuson, Å. C. (2009). Prediction of solubility curves and melting properties of organic and pharmaceutical compounds. *European Journal of Pharmaceutical Sciences*, *36*(2–3). https://doi.org/10.1016/j.ejps.2008.10.009

Shiri, P., Lai, V., Zepel, T., Griffin, D., Reifman, J., Clark, S., Grunert, S., Yunker, L. P. E., Steiner, S., Situ, H., Yang, F., Prieto, P. L., & Hein, J. E. (2021). Automated solubility screening platform using computer vision. *IScience*, *24*(3). https://doi.org/10.1016/j.isci.2021.102176

Tan, J. S., Hilden, L. R., & Merritt, J. M. (2019). Applications of In Silico Solvent Screening and an Interactive Web-Based Portal for Pharmaceutical Crystallization Process Development. *Journal of Pharmaceutical Sciences*, *108*(8). https://doi.org/10.1016/j.xphs.2019.03.013

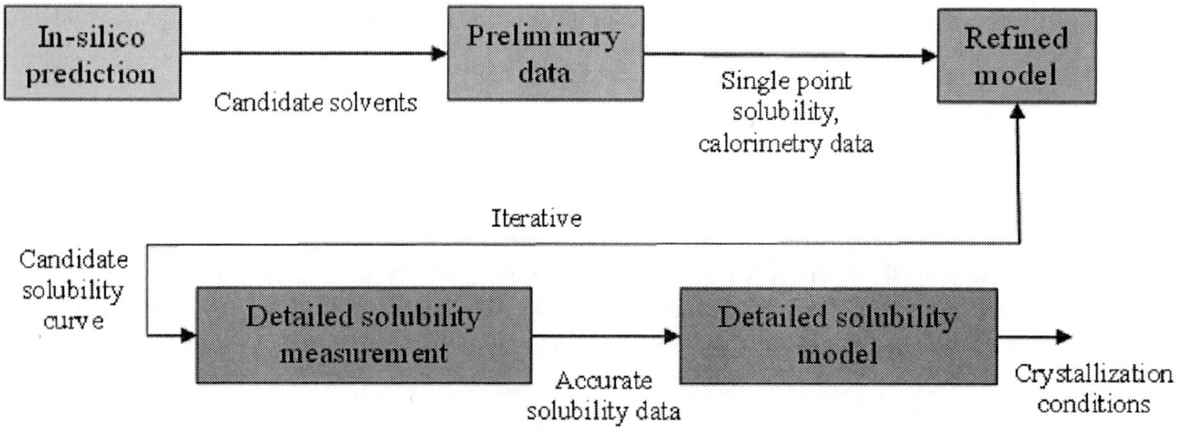

Figure 1. Takeda Workflow for solvent screening and solubility measurement.

655d Fully Automated and High-Throughput Process Development and Optimization for Solid Oral Dosage Forms

Thomas Brinz

INTRODUCTION

Quality by design is a systematic approach for the development of pharmaceutical products, to enhance process understanding and finally, a better product quality. The understanding of the process, based on sound science and quality risk management makes a decisive contribution to a successful process capability and validation. According to the International Council for Harmonization (ICH), the Quality by Design (QbD) principle requires an analysis of the functional relationships between the material properties, the process parameters, and the critical quality features of the final product.
Implementing QbD often requires investment in resources such as time, expertise, and technology. Individuals or organizations with limited resources may find it challenging to adopt QbD practices, especially if they perceive the initial costs and effort as prohibitive. Establishing QbD may require significant changes to existing practices. The presented Automated process development refers to the use of automation and advanced technologies to streamline and accelerate the development of manufacturing processes. It offers numerous benefits, including speed, efficiency, data accuracy, resource optimization, complexity handling, iterative optimization, risk reduction, scalability, adaptability and flexibility.

Automated Process development

When a new formulation, capsule filling or tablet compression process has to go into operation, the primary goal in the development should not be only, to find the appropriate process parameters that enable filling within the specified process limits. Rather, it is crucial to investigate the entire range of process parameters globally in order to understand their influences and to operate the process at a robust and optimized operating point. This ensures safe and stable production.

Various influences can affect the result of a powder formulation, tablet compression or capsule filling process. These are, among others, the directly adjustable process parameters on the machine like the machine speed. Other process parameters are associated with a machine conversion and take a greater effort to change (e.g., the change between 2- and 3-paddle-feeder).

Finally, the material properties must be mentioned: Experience has shown that these have the greatest influence on the process. Each pharmaceutical substance has different chemical and physical properties that have a considerable influence on filling behavior.

A capsule filling or compression process can be described as a black box (Fig. 1). On the input side are the process parameters and the material properties that are fed into the capsule filling or compression process. On the output side is the result of the solid dosage form, described from the target values for the final product, also called Critical Quality Attributes (CQAs). For a capsule filling process, these include e.g., the average weight, the uniformity of the dimensions and the composition.

The decisive factor is now what happens inside the black box during the process, determined by the critical material attributes (CMAs) and the critical process parameters (CPPs). For the process commissioning, it is important to analyze all influences and interactions on the critical quality attributes.

There is no doubt that the performance of machine tests is associated with a prominent level of effort: Product is required for each test run, the tests must be conducted and then evaluated. To reduce this effort, the automated process development (APD) method can be used. Automated testing starts with statistically optimized test plans and carries them out automatically with reduced powder consumption on the corresponding system or machine.

Based on the experiments and the results, a model of the process can be created. This model describes the effects and interactions that are required for understanding the process and allows a prediction of the CQAs. The execution with the material on the real machine will later offer the advantage that the production process behaves in the same way as in the analysis. The process becomes transparent and can be validated, also better control strategies can be established.

The use of statistical design of experiments (DoE) offers significant added value compared to classical design of experiments. Here, the number of necessary experiments is considerably reduced in order to achieve the same informative value compared to an empirical experimental design. Furthermore, the effort required to set the process parameters on the machine is reduced. A high degree of automation during all processes enables the required settings to be made within a few seconds without manual handling steps and material loss. A higher-level controller automatically takes over the tests from the test plan and sets the process parameters automatically.

Automated Blending and fluid bed Granulation process

R&D equipment designed for semi-continuous manufacturing can be used to conduct fully automated and DOE-planned development of a solid dosage form with variation of critical process parameters in powder dosing, blending and fluidized bed granulation. Beside variation of API-content, different grades of excipients and content process parameters with major impact like mixing speed and time, inlet air temperature, inlet air volume rate, spray rate and spray air pressure are investigated in a short period of time.

Capsule Filling process

In the presentation a capsule filler with the automated adjustment of al critical process parameters, an advanced sensor technology to control material and process conditions and process monitoring and the automated change-over to the next parameter set will be presented (fig.2).

Automated process development can be used in process optimization. Combined with lower material consumption, the correlation of process efficiency and quality can be evaluated in 2 hours, compared to 2 weeks of standard manual work.

Interactions of process parameters make it often complex to optimize machine processes. In different tests we show, that different and sometimes totally, opposite process parameters can lead to a robust process. Examples of typical results are shown in figure 2.

Automated Process Development in tablet compression

The feeder of a rotary tablet press is usually operated with either 2 paddles (2-paddle feeder) or 3 paddles (3-paddle feeder). The studies in this article show that the mode of operation is a decisive and statistically significant influencing factor in the determination of parameters with the aim of reducing scatter in tablet weight and tablet hardness. A new automation solution for tablet presses will be presented, which enables a changeover from 2-paddle operation (2-paddle feeder) to 3-paddle operation (3-paddle feeder) as well as between diverse types of paddle shapes without manual intervention and machine stops. The comparison of the time required to a manual changeover provides an overview of the potential of the automated feeder for prompt and meaningful predictions for the optimal parameter settings on the tablet press. Different studies with poor flowing materials, identification of over-lubrication and capping effects will be shown.

Fully Automated and DoE-Based Development of an Oral Solid Dosage Form

The aim of this study is to demonstrate the potential of fully automated and DoE-based development of oral solid dosage forms (tablets). The combination of design-of-experiments and automated execution of the individual process steps offers a new potential to accelerate the development process and to integrate more Quality-by-Design.

- Planning of all trials using Design-of-Experiments
- Automated execution of all experiments and analysis of the results

By automating all development steps (granulation, compression and quality analysis) a high throughput and short development time were achieved. 26 different fluid bed granulation experiments were performed, more than 260 tablet batches were prepared, each containing more than 400 tablets. All tablets were evaluated on a new tablet analysis tool, measuring weight, dimension and API content. Also, the sophisticated investigation of interactions between formulation process parameters with the quality attributes of the compression step drives the understanding of the final manufacturing process and quality to a new level. The complete development process, from the powder to the final tablet, including determination and analysis of Critical Process Parameters and Critical Quality Attributes was performed in 7 days.

By combining DoE and automation it is possible to screen a larger parameter space, detect interactions of parameters, and optimize tablet quality (Quality by Design). Due to automation, the amount of material required is less than with manual testing and the intermediate change of the various modules such as 2- and 3-paddle-feeder. The method is transferrable to all kinds of oral solid dosage processes and products. It also can be used in the development of batch and continuous manufacturing products.

Conclusion

Automated process development combines Design of Experiments, automated execution of tests with the integrated measurement of quality attributes. Lower material consumption, more efficient process optimization and identification of critical process parameters make a pronounced contribution to reducing costs, improving quality and efficiency.

FIG 1: PRINCIPLE OF AUTOMATED PROCESS DEVELOPMENT

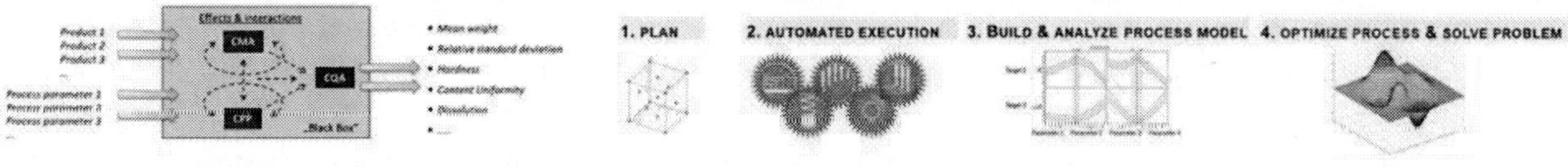

FIG 2: PROCESS MODEL OF FLUID BED GRANULATION, TABLET COMPRESSION AND CAPSULE FILLING

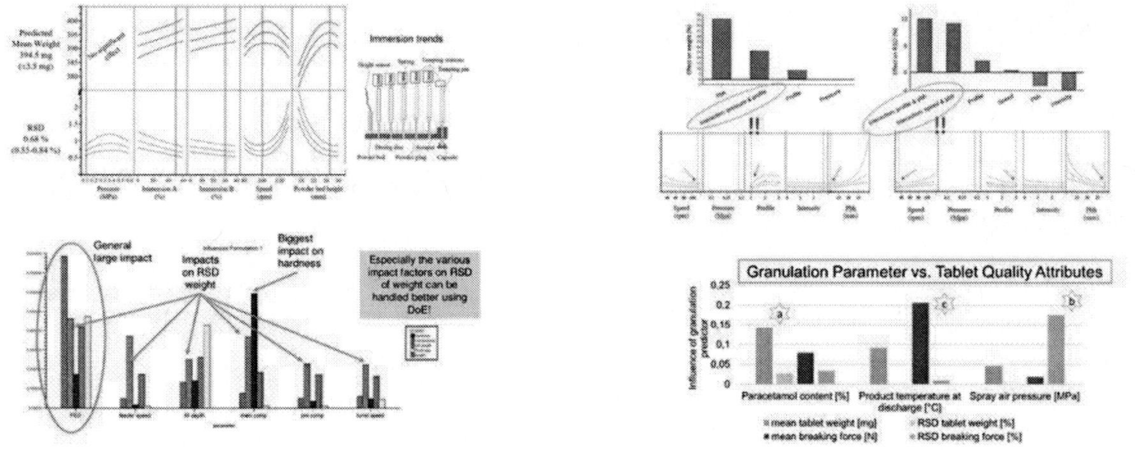

655e Predicting Granule Flowability Changes Due to Friability Under Low Shear Conditions before Tablet Production

Indu Muthancheri, Abhishek Paul, Mario Rousselin

Powders are often granulated to improve their flow behavior before tableting. Granule friability is the tendency of granules to lose components due to abrasion, friction, or mechanical shock. The friability of granules becomes crucial when handling granules between granulation and tableting operations. It is affected by the strength of granules, the mechanical properties (elastic, plastic, brittle nature) of the material, and the size and porosity of granules. Friable granules (granules with low strength) lead to bimodal particle size distribution and fines generation [1], affecting powder flowability and potentially causing segregation in the final blend during die-filling. Fine particles are often re-granulated to improve yield, but studies have shown content non-uniformity issues related to the recycling of fines [2]. Considering flowability challenges, the risk of segregation, and tablet content non-uniformity, it is of interest to study the effect of granule friability on the flowability of granules.

This work presents a case study to systematically integrate granule friability and flowability models into the drug product development workflow. As a first objective, the granule friability during the powder-feeding operation is predicted using a population balance model framework. Granule chipping breakage mechanism is considered the dominant mechanism in the powder feeder, similar to the low shear conditions of a conveying element in twin screw granulator [3]. The granule strength-dependent breakage function proposed by Wang et. al. [4] is utilized. Subsequently, the flowability of friable granules is studied as the second objective. The granule size distribution is used to predict the flowability of granules using a modification to the rheology model suggested in the literature [5]. This work will highlight a practical methodology for adapting a mechanistic model to improve process understanding during process optimization studies during late-stage drug product development.

[1] Metta, Nirupaplava, et al. 'Model development and prediction of particle size distribution, density and friability of a comilling operation in a continuous pharmaceutical manufacturing process.' International journal of pharmaceutics 549.1-2 (2018): 271-282.

[2] Sheskey, Paul J., et al. 'Use of roller compaction in the preparation of controlled-release hydrophilic matrix tablets containing methylcellulose and hydroxypropyl methylcellulose polymers.' Pharmaceutical technology 18.9 (1994): 132-132.

[3] Dhenge, Ranjit M., et al. 'Twin screw granulation using conveying screws: Effects of viscosity of granulation liquids and flow of powders.' Powder Technology 238 (2013): 77-90.

[4] Wang, Li Ge, et al. 'A breakage kernel for use in population balance modelling of twin screw granulation.' Powder technology 363 (2020): 525-540.

[5] Roy, Sudeshna, Stefan Luding, and Thomas Weinhart. 'A general (ized) local rheology for wet granular materials.' New journal of physics 19.4 (2017): 043014.

Author Disclosure: All authors are Sanofi employees and may hold shares and/or stock options in the company.

655g Fast-Tracking Process Optimization with Small Data and Machine Learning

Daniela Blanco, Ethan Tenison

Pharmaceutical process design and optimization is a time-consuming and complex task. Data-driven machine learning (ML) can accelerate the process by predicting behavioral trends, variable effects, and optimal conditions. However, conventional ML algorithms often require extensive datasets, posing a resource-intensive challenge in experimental data collection.

To address this bottleneck, we introduce an enhanced Bayesian optimization (BO) approach, enabling a reaction-agnostic pathway for designing and implementing intelligent experimental campaigns. SunthetiesML is an accessible online ML platform tailored for researchers without coding or ML expertise. The approach demonstrates compelling returns on material and experimental efficiency, as well as performance gains against a competitive baseline.

SunthetiesML is a versatile technology that allows numeric, discrete, and mixed-integer optimization problems with up to 20 input parameters, further facilitating bounded-target, multi-objective, and constrained-input optimizations. We will show case studies of SunthetiesML with industrial partners showcasing accelerated formulation optimization, process characterizations, and process development efforts in chemocatalytic reactions, biocatalytic cascades, crystallizations, in-vitro mRNA transcription processes, and more.

The in-lab validation of SunthetiesML convincingly demonstrates impressive returns on material efficiency, with up to a 75% reduction in the use of expensive or complex reagents. Experimental efficiencies enabling 9-12% increases in previously-optimized yields with a 2-32X reduction in optimization experiments.

The insights gained from this work redefine the landscape of reaction engineering, process development, and optimization, while simultaneously lowering barriers to the adoption of new chemical technologies.

218a Modeling and Experimental Investigation of Polymorphic Imatinib Mesylate Crystallization in Batch and Continuous Configurations

Hemalatha Kilari, Ilke Akturk, Zoltan Nagy

Interest in polymorphism has increased markedly in pharmaceutical industry since two decades. The emphasis on polymorphism is due to various factors such as the polymorphic form is known to have direct impact on drug substance processability, drug product manufacturability and product quality or performance such as stability, dissolution and bioavailability [1]. Polymorphic screening in pharmaceutical industry is important not only to provide necessary knowledge for process development but also to deal with intellectual property issues or patent regulations [2]. Imatinib is a ground-breaking first-generation tyrosine kinase inhibitor, and has revolutionized the paradigm of targeted cancer therapy since its inception. It is used in the frontline treatments against chronic myeloid leukemia (CML). Its efficacy was also proven in treatment of gastrointestinal stromal tumors, and also in other malignancies that involve expression of a tyrosine kinase [3]. Due to its potent bioavailability, several flow synthesis methods have been developed for imatinib and its salts. Imatinib-free base is insoluble in water, hence its derivative, Imatinib mesylate salt is used in oral drug formulations and is very highly soluble in aqueous medium of pH < 5.5. Imatinib is classified under biopharmaceutical classification system (BCS) as BCS-class I drug by FDA (high permeability and high dissolution) [4]. Imatinib mesylate displays polymorphism, with both metastable α-form and stable β-form being utilized in commercial drug formulations. Consequently, the polymorphic crystallization of imatinib mesylate has garnered significant interest.

In our work, the effect of process variables such as temperature, solvent, seeding and batch time/residence time on the final polymorphic form produced is studied. As the metastable form has high solubility in methanol compared to the stable form, recrystallizing the stable form from the saturated solution of the metastable form proved beneficial in-terms of creating high supersaturations with respect to stable form and also provide high yields and volume productivity. Batch crystallization experiments revealed that high seeded experiments, seeded with metastable form and performing cooling crystallization with shorter batch times favored the crystallization of metastable form and slower cooling rates with longer holding times favored crystallization of stable form. The possibility of producing various forms selectively by varying residence times and seeding is explored in continuous configuration with a cascade of mixed suspension and mixed product removal crystallizers (MSMPR's). However, as the main objective is to establish a continuous crystallization model for imatinib mesylate, we combine the knowledge of process analytical technology tools (PAT) (inline infrared spectroscopy, Raman spectroscopy, imaging) and offline characterization (PXRD, particle size distribution) with population balance modeling to investigate the crystallization of stable form from saturated solution of metastable form in a continuous manner at steady state. The solvent mediated polymorphic transformation of α-form form to stable β-form is more suitable for crystallization as β-form crystals are cubic shaped and have better flowability and compaction characteristics compared to needle shaped crystals of α-form. The outcomes illustrated effective model establishment for both batch and mixed-suspension-mixed-product-removal (MSMPR) crystallization configurations. This experimentally validated model, serves as a digital twin of the process which is used to optimize the MSMPR process to produce larger crystals with narrower distributions.

References

1. Chambi, J. T., Fandaruff, C., & Cuffini, S. L. (2024). Identification and quantification

techniques of polymorphic forms-A review. *Journal of Pharmaceutical and Biomedical Analysis*, 116038.

2. Komai, T., Togo, T., Uchiyama, H., Kadota, K., & Tozuka, Y. (2024). Kinetics of crystal transformation of imatinib mesylate during formulation process: Selection of dry granulation in pharmaceutical process. *Powder Technology*, 119570.
3. Imran, M., Asdaq, S. M. B., Khan, S. A., Unnikrishnan Meenakshi, D., Alamri, A. S., Alsanie, W. F., & Bakht, M. A. (2021). Innovations and patent trends in the development of USFDA approved protein kinase inhibitors in the last two decades. *Pharmaceuticals*, *14*(8), 710.
4. Maddeboina, K., Yada, B., Kumari, S., McHale, C., Pal, D., & Durden, D. L. (2024). Recent advances in multitarget-directed ligands via in silico drug discovery. *Drug Discovery Today*, 103904.

218b Integrating Functional Principal Component Analysis with Data-Rich Experimentation for Enhanced Drug Substance Development

Brian Wyvratt, Jonathan McMullen, Cynthia Hong, Akasha Purohit

The combination of laboratory automation and data-rich experimentation has led to significant improvements in experimental efficiency, reproducibility, and throughput compared to conventional methods. These transformative technologies have proven particularly valuable in drug substance process characterization studies where the ability to rapidly generate volumes of reaction data to maximize understanding for ever-increasing regulatory expectations is of paramount importance. These highly dense datasets offer significant potential when coupled with data-driven modeling to quantify reaction dynamics and sensitivities for process knowledge and optimization. To ensure that this analysis is applied throughout drug substance reaction development, identifying a data-driven modeling approach capable of describing diverse reaction trends is of critical need. In this study, functional principal component analysis for data-driven reaction modeling was performed to highlight its applicability for drug substance development. To demonstrate this methodology, we employed automated reactor and sampling technologies in the process characterization studies of a heterogeneous fluorination reaction using gaseous trimethylamine. These data-rich experiments were structured according to a 2^4 full factorial design of experiment, each comprising of at least 12 reaction time samples. By applying functional principal component analysis to various reaction responses, the optimal design space for manufacturing operations was easily identified.

218c Data-Driven Process Decisions for Optimization and Robustness: Leveraging Data-Rich Experimentation and Process Modeling

Jonathan McMullen, Victoria Zhang, Kotoe Abe, Kevin Sirk, Sean Dubina, Victor Schultz, Arnav Malkani, Kaitlyn Brinza, Timothy Wright, Morgan Crawford, Carmen Chen, Andrea Angulo

Process development and manufacturing of drug substance material often requires an extensive and detailed experimental strategy to ensure process robustness, successful scale-up, and a suitable control strategy to provide quality active pharmaceutical ingredient (API). However, the burden of generating sufficient data to meet these goals is exacerbated by accelerated project timelines, limited resources, and competing priorities. Consequently, there is a growing need to develop novel experimental approaches and leverage process models to quickly build process understanding and apply the knowledge to make meaningful process improvements.

In this work, we describe methodologies that leveraged automation technologies and data-rich experimentation strategies on multiple synthetic steps in the manufacturing route of an API. By integrating these strategies into the process development of each unit operation, the process knowledge gained from each experiment was maximized in a resource sparring manner. Reaction process fingerprints were collected using process analytical technologies for in situ monitoring and automated sampling technology for off-line chromatographic analysis. The time-series data that was collected from these experiments was used to build mechanistic models and data-driven models to gain critical process insight, optimize operating conditions, and ensure quality drug substance. As needed, empirical relationships were used to scale these laboratory observations to the production environment. Ultimately, the methodology employed in this work enabled a fast and efficient process development to achieve program goals on an accelerated timeline.

218d A Simple Method for Incorporating Prior Parameter Information during Model-Based D-Optimal Design of Experiments in Pharmaceutical Production

Lauren Gibson, Fei Liu, Kimberley B. McAuley

Mathematical models are important for process design, control, and optimization for chemical, biological and pharmaceutical systems.[1] These fundamental models often contain many unknown parameters, such as kinetic rate constants and activation energies, that require estimation from experimental data. Typically, only limited data are available for parameter estimation, making it difficult to estimate all the unknown parameters reliably.[2] When modelers wish to obtain more-accurate estimates for model parameters, they may choose to perform new experiments to obtain additional information. Performing new experiments can be expensive and time-consuming, so researchers want to carefully select operating conditions so that as much information as possible is obtained from the new experimental runs. Model-based design of experiments (MBDoE) is an efficient way to plan experiments that lead to accurate parameter estimates and improved reliability of the model.[3]

For nonlinear models with many unknown parameters, computation of MBDoE objective functions is a challenge due to a singular Fisher information matrix (FIM).[4,5,6] Several methods have been developed to overcome the problem of a singular FIM, such as parameter-subset selection[4] and Bayesian approaches.[5] Unfortunately, closed-form objective functions for Bayesian MBDoE are difficult to obtain.[7] Therefore, sampling-based approaches, such as Markov chain Monte Carlo (MCMC) methods, are typically used for Bayesian MBDoE.[8,9] However, these methods tend to be complex to understand and require excessive computation for large models, such as many chemical and pharmaceutical process models.[8]

In this study, we propose a simplified linearization-based Bayesian D-optimal objective function for MBDoE in multi-response nonlinear models. The proposed objective function will be useful for modelers who want to design new experiments to obtain better parameter estimates for nonlinear pharmaceutical process models with many unknown parameters. We also developed a new method to test the effectiveness of experimental design methods for nonlinear models in a D-optimal sense. This new method relies on the computation of the volumes of minimum volume ellipsoids (MVEs) that are centered at the true parameter values. The proposed MVE methodology will be helpful to future researchers who want to compare the effectiveness of different experimental design methods aimed at achieving accurate parameter estimates and model predictions used for mapping of permissible operating regions.

A pharmaceutical case study is used to illustrate and test the proposed Bayesian D-optimal objective function for designing new experiments. This case study involves a nonlinear dynamic model of a batch reactor with Michaelis-Menten kinetics, which is used for the production of a pharmaceutical product.[10] Initially, only four replicate runs are available to estimate the 14 unknown parameters (kinetic rate constants and activation energies). The proposed simple Bayesian D-optimal objective function is used to design two new experiments in a variety of ways: two experiments one-at-a-time and two experiments at-the-same-time. We confirm that, as new D-optimal experiments are included in parameter estimation, the MVE of the resulting parameter estimates gets smaller. We also use a similar Bayesian A-optimal objective function, developed in a previous study[5], to design new A-optimal experiments. We show that estimates from D-optimal experiments have a much smaller MVE than corresponding estimates from A-optimal experiments or randomly-selected corner-point experiments. As a result, we confirm that the proposed Bayesian D-optimal objective function is effective in a D-optimal sense. We anticipate that the proposed Bayesian D-optimal methodology will be useful to a wide variety of modelers who develop pharmaceutical models with a large number of parameters, because of its computational simplicity relative to MCMC-based methods. When planning new experiments, future modelers will be able to readily account for prior beliefs about plausible parameter values, the structure of their model equations, and old data that may be available.

References

1. Maria G. A review of algorithms and trends in kinetic model identification for chemical and biochemical systems. *Chemical and Biochemical Engineering Quarterly*. 2004;18:195-222.
2. McLean KAP, McAuley KB. Mathematical modelling of chemical processes – obtaining the best model predictions and parameter estimates using identifiability and estimability procedures. *The Canadian Journal of Chemical Engineering*. 2012;90:351-366.
3. Box GE, Lucas HL. Design of experiments in non-linear situations. *Biometrika*. 1959;46:77-90.
4. Thompson DE, McAuley KB, McLellan PJ. Design of optimal sequential experiments to improve model predictions from a polyethylene molecular weight distribution model. *Macromolecular Reaction Engineering*. 2010;4:73-85.
5. Shahmohammadi A, McAuley KB. Using prior parameter knowledge in model-based design of experiments for pharmaceutical production. *AIChE Journal*. 2020;66:e17021.
6. Walz O, Djelassi H, Mitsos A. Optimal experimental design for optimal process design: A trilevel optimization formulation. *AIChE Journal*. 2020;66:e16788.
7. Chaloner K, Verdinelli I. Bayesian experimental design: A review. *Statistical Science*. 1995;10:273-304.
8. Ryan EG, Drovandi CC, Pettitt AN. Simulation-based fully Bayesian experimental design for mixed effects models. *Computational Statistics and Data Analysis*. 2015;92:26-39.
9. Kalyanaraman J, Kawajiri Y, Realff MJ. Bayesian design of experiments for adsorption isotherm modeling. *Computers and Chemical Engineering*. 2020;135:106774.
10. Domagalski NR, Mack BC, Tabora JE. Analysis of design of experiments with dynamic responses. *Organic Process Research & Development*. 2015;19:1667-1682.

218e Improving Our Understanding of Nitrosamine Formation in Pharmaceutics: Computational and Experimental Approaches

Ayse Eren, Carla Luciani

With the pharmaceutical industry increasingly coming under scrutiny for the presence of nitrosamines in their products,[1] it is important for industry to create tools to better understand nitrosamine formation mechanisms, prevalence, and methods for prevention and removal.

Amines and nitrosating agents, the precursors of nitrosamines, are ubiquitous, with traces found in seemingly innocuous sources.[2] Given the prevalence of amines and nitrogen-containing compounds in various pharmaceutical processes, tools for conducting risk assessments, including kinetic models that describe the nitrosation reaction processes have been used by researchers for a while.[2–4] Leveraging these tools allows a quantitative evaluation of the risks associated with nitrosamine formation to be conducted and effective design of control measures implemented, as needed.

In this work, we investigate the opportunity to enhance existing kinetic models by incorporating additional phenomena, such as equilibrium equations, nitrosating agent decomposition reactions, along with the dissociation behavior or ions that facilitate nitrosation reactions.

Experiments were designed to evaluate nitrosation reactions in solution phase. In this work, we developed a comprehensive mathematical model to capture the behavior of typical pharmaceutical processes and to carry out a realistic risk assessment of nitrosamine formation. The model includes the dissociation equations of halides, the impact of solubility of amines, and other equilibria to enhance the already established solution-phase reaction model.[2] Our simulation results shine light on potential strategies that can be used to mitigate nitrosamine formation risk.

References:

(1) R. Nudelman, G. Kocks, B. Mouton, D.J. Ponting, J. Schlingemann, S. Simon, G.F. Smith, A. Teasdale, A.L. Werner, The Nitrosamine “Saga”: Lessons Learned from Five Years of Scrutiny, Org Process Res Dev 27 (2023) 1719–1735. https://doi.org/10.1021/acs.oprd.3c00100

(2) Ashworth, I. W.; Dirat, O.; Teasdale, A.; Whiting, M. Potential for the Formation of N-Nitrosamines during the Manufacture of Active Pharmaceutical Ingredients: An Assessment of the Risk Posed by Trace Nitrite in Water. *Org Process Res Dev* **2020**, *24* (9), 1629–1646. https://doi.org/10.1021/acs.oprd.0c00224 .

(3) Rayson, M. S.; MacKie, J. C.; Kennedy, E. M.; Dlugogorski, B. Z. Accurate Rate Constants for Decomposition of Aqueous Nitrous Acid. *Inorg Chem* **2012**, *51* (4), 2178–2185. https://doi.org/10.1021/ic202081z .

(4) Licht, W. R.; Deen, W. M. Theoretical Model for Predicting Rates of Nitrosamine and Nitrosamide Formation in the Human Stomach. *Carclnogenesis* **1988**, *9* (12), 2227–2237. https://doi.org/https://doi.org/10.1093/carcin/9.12.2227 .

218f Advancing Drug Product Stability Prediction Via Integrating Data Science, Decision Tree, and Kinetic Modeling for Continuous Improvements

Nick Yeh, Sedra Khan, Evgenyi Shalaev, Jie Wang, vijay Kumar Velagala, Zhibin Tan, Chetankumar Malepati, Sherwin Shang

Pharmaceutical drug product (DP) early development can offer an initial prediction of DP stability with limited data of lab scales for the kinetic modeling exploration. By continuing the DP stability study through further development at pilot scales and manufacturing scales and complementing it with additional data of kinetic modeling analysis, the accuracy and reliability of DP stability prediction can be improved.

The objective of this study is to advance drug product stability prediction beyond just using lab generated data. The modeling process is proposed to include, but not limited to employing a) data science methodology to analyze available stability data, b) decision tree analysis to reveal attributes-performance classification for trend analysis, and c) integrating both of them into kinetic modeling with accessible manufacturing scale data, to develop drug product's real-life scenarios for innovations and improvements of the DP stability models for modeling and prediction.

Once the stability kinetic modeling results from the early lab stage trends are in alignment with the observed trends at the manufacturing scale, it signifies that the DP stability can be strongly dependent on few critical parameters, as observed at the lab scales. This indicates that there are no additional key attributes from the scale up and tech transfer that have an impact on drug product stability. In other words, the kinetic modeling projection from the early lab scale development can act as a representative for scaled manufacturing system.

In contrast, the misalignment observations will suggest from a system-level that there should be additional attributes impacting DP stability. These are particularly related to the equipment and process scale up to pilot and manufacturing scales. The kinetic modeling from accessible manufacturing scale data can guide additional attributes impacting the DP stability. Accordingly, the integrated methodology of data science, decision tree, and kinetic modeling can enhance the probability of success for stability prediction. This can be achieved with the newly identified additional key attributes, either from scale up or tech transfer to improve the manufacturing process control, and thus to enhance DP stability and modeling prediction.

218g Bayesian Optimization Applied to Chemical Reaction Optimization: Human-in-Loop and Cdmo Perspective

Rudi Oliveira, Ines Rocha, Saúl Silva, Pedro Valente

Development of a chemical synthesis involves optimization of multiple parameters, such as: substrates; catalysts; reagents; solvents; concentrations; and temperatures, among other variables. Due to timeline and cost restrictions, it is not possible to study the full range of process parameters and only a limited number of experiments are considered to find optimal conditions. Scientists typically draw upon their accumulated knowledge and available literature, to identify key parameters that most significantly influence the success of a reaction at scale. To complement this Human effort, there has been significant advances in automated optimization techniques, also referred to as Algorithm Process Optimization (APO).[1] Among these, Bayesian Optimization (BO) has proven to be particularly effective, often surpassing the acumen of human experts and other advanced optimization algorithms.[2,3] It alternates between exploring uncertain regions of an operational space and exploiting known data. This method has the advantage of being applicable to a wide range of parameter spaces, including those with complex, parameterized domains, and can design multiple parallel experiments.

To illustrate the performance and robustness of different process optimization strategies, we conducted a "virtual lab" case study based on the work developed by Shields et al..[3] This study was designed around a high throughput experimentation (HTE) database, complemented with a calculated dataset for level of impurities and a green score, aiming to add real-world relevance in the complex task of optimizing a process for performance and sustainability. The case-study presented allows to benchmark traditional human SME with BO in the effectiveness and efficiency in optimizing a given chemical reaction, but also to shed light on how experts from different backgrounds and expertise levels interact with a BO tool.

An initial comparative analysis of the optimization data (**Figure 1**) shows that larger green score values occur for lower reaction yields, even though it linearly increases with yields when all other factors are kept constant. This reflects the inherent trade-offs faced in optimizing for both reaction efficiency and sustainability. All but one scientist reached at least one optimum condition of the Pareto Front, i. e. achieving an average between the Yield and Green Score within the top four possible values conditioned on impurities being within the specified limit.

Clustering the design space and results in a t-distributed Stochastic Neighbor Embedding (t-SNE) visualization,[4] allowed to map the individual optimization trajectories of different scientists towards finding the Pareto optima. Even though the optimization outcome was similar in some cases, the strategies taken were starkly different, even among scientists using Bayesian optimization which can lead the optimization on its own without Human interference. This demonstrates that Human input can either improve or deteriorate performance of the algorithm. As such, there is the need for better human-in-the-loop strategies that can combine the algorithms with human expertise, streamlining the path through the inherent complexities of organic synthesis and unlocking new possibilities in sustainable process development.

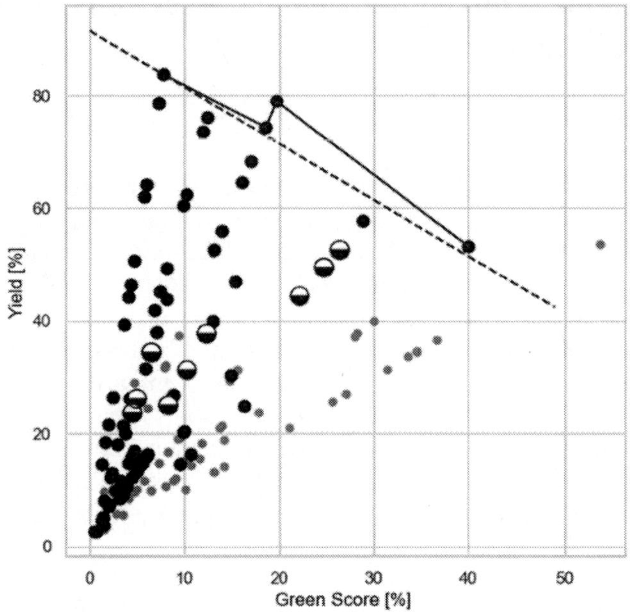

Figure 1. Overview of the reaction yield and green score for each of the reaction conditions in the design space of the current case-study. Black (•) and grey (•) markers represent data points that either meet or not the specified impurity criteria, respectively. Initial experimental data package is represented as half-filled markers (◒). The pareto front is shown as the filled line (—), whereas the dashed line marks the region for the top four experimental outputs (--).

218h Adequacy Testing and Lifecycle Management for a Soft-Sensor Based on State-Estimation Approaches. Case Study: Fluid Bed Granulation

Salvador García-Muñoz, Pedro De Azevedo Delou

The utilization of deterministic process models for pharmaceutical development is an established practice across the pharmaceutical research and development sector. And despite the current trends in the utilization of artificial intelligence, the application of transport, conservation, and chemistry laws in the derivation of mathematical models (when possible) remains the golden standard. Because models are fit for use, the assumptions and parametrization of these need to be in line with the intended use. And since even the simplest model can easily become overparametrized the practitioner should be cautious of carefully managing physical detail that becomes non-estimable from the measurements and experiments available. Tools like estimability analysis can provide the modeler with useful information about the extent of parametrization that a model can have, based on the available data. Low- and medium-impact deterministic models are then commonly utilized to define operating conditions and ranges for commercial manufacture.

The lifecycle of a deterministic model normally ends at the tech-transfer stage due to the lifecycle requirements for a model-based solution. ICH calls for a model maintenance approach that mostly focuses on ensuring that the accuracy and adequacy of the model is kept with respect to the product lifecycle in manufacturing. These approaches are however insufficient when a model is deployed as a component of a more complex exercise such as that of a state estimator. One of the advantages of using a state estimation approach is the potential to produce accurate estimates for a material attribute using process data only, avoiding the need for an analytical instrument (such as a spectrometer).

A state estimator, such as the Extended Kalman filter (EKF), uses a process model in combination with the available output measurements on an optimization algorithm to update the model state variables in real time such that the produced estimation presents an optimal compromise between the model and measurement uncertainties. When model parameters are estimated simultaneously with state variables, the state estimation optimally adapts the model to resembles the operation accounting for the effect of the unmeasured disturbances. The state estimator then uses measurements from model inputs and measured model outputs to reconstruct all states and parameters including any observable un-measured states (such as an unmeasured material attribute). This calculation does not rely exclusively on the accuracy of a model for a specific set of parameters; but relies more on the model's structure and flexibility to mimic the physics of the process, irrespective of the specific magnitude of the states within the space where the assumptions hold. As such, model-specific metrics to assess the performance of a state estimator are insufficient and inappropriate to regulate the lifecycle of a state estimator.

In this talk, we will present a set of diagnostics and their use in a proposed life cycle for a soft sensor that is built using state estimation. We illustrate these concepts in the development of a Loss on Drying (LOD) soft sensor using an EKF for a fluid bed granulation.

694a Enhancing Drug Processibility through Improved Dry Coating Techniques: A Comparative Study of Lab and Pilot Scale Methods

Siddharth Tripathi, Sebastian Escotet-Espinoza, James Tarabokija, Jerry Klinzing, James DiNunzio, Rajesh Dave

Dry coating of pharmaceutical powders is a promising technique that can enhance their bulk properties and downstream processibility during tablet manufacturing. By applying a thin, sparse layer of nano-silica onto the surface of the powders, various beneficial effects such as improved flowability, reduced cohesion, and enhanced tabletability can be achieved. This process can be performed using different dry coating techniques, including acoustic mixing, milling, and diffusive blending. Comprehensive previous studies have investigated the efficacy of nano-silica dry coating and emphasized the importance of optimized coating parameters and specific silica types. However, further exploration is needed to understand the implementation of dry coating at different scales and its impact on downstream drug product attributes. This work focuses on evaluating the benefits of dry coating in four industrially-relevant investigations: (1) comparing the addition of nano-silica using surface area coverage (SAC) and weight percent-based approaches, (2) assessing the impact of scale-up on dry coating operations, (3) examining the benefits of dry coating on feedability of powders, and (4) investigating the enhancements of dry coating on tabletability and sticking.

Determining the appropriate amount of nano-silica needed for dry coating is critical to obtain desired processability enhancements. This study compared the traditional 1 wt.% silica addition methodology with a SAC-based approach at the lab-scale using a LabRAM acoustic mixer. A lab-scale study evaluated six materials consisting of three APIs and three excipients with particle sizes ranging from 2-40μm (d50). It was observed that dry coated powders are distinguishable and well characterized when such tests are employed which require external consolidation like FFC. Hence flow function coefficient (FFC) was found to be a reliable indicator of flowability enhancement after coating. The SAC-based approach was shown to be superior to the arbitrary 1 wt.% silica blending approach, as materials coated to 50% SAC with nano-silica content ranging from 0.5-4 wt.% displayed better flow performance. However, practical challenges were identified in determining the varying composition of nano-silica during industrial drug product development. Recommendations were provided for specific silica amounts tailored to host particle size ranges, considering the variability in size and density of host and guest particles. This workflow establishes an industrially feasible approach that provides weight percent ranges based on host particle attributes, resulting in significant flow improvements.

For scalability analysis from lab to pilot scale, a COMIL-U10 was utilized as a continuous dry coating device, focusing on three materials that required 1 wt.% silica for adequate coverage. The study aimed to investigate the efficacy of silicon dioxide dry coating techniques using COMIL-U10 and lab scale LabRAM to enhance bulk properties and evaluate their impact on downstream API processibility for tableting. Two silica variants, Cab-O-Sil M5P (hydrophobic) and Aerosil R972P (hydrophilic), were examined in terms of bulk property improvement and downstream processibility. The results indicated that the performance of pilot-scale equipment was not as efficient as lab-scale equipment, although it still provided measurable improvement in powder flow properties. Feeding tests conducted on pilot-scale coated materials (using the COMIL-U10) showed excellent improvement in feeding uniformity.

Lastly, the processability benefits of dry coating were evaluated during the tableting stage. Two studies were performed: a tabletability assessment at varying compression speeds and a sticking assessment. Compaction trials were conducted with varying drug loads of 10%, 30%, and 60% to compare the dry coating effectiveness of LabRAM and COMIL with the traditional 1 wt.% silica blending approach. The analysis revealed significant impacts on tablet attributes, with the highest tensile strength observed in LabRAM-coated API cases, intermediate in COMIL-coated cases, and the lowest in traditional silica blended formulation tablets. This highlights the critical role of shear in silica dispersion and coating effectiveness. Moreover, both LabRAM and COMIL facilitated high-speed tableting at higher drug loads (30% and 60%), while the silica added cases failed beyond a drug load of 10%. Importantly, the enhanced tablet properties observed in LabRAM and COMIL coated API formulation cases were achieved with significantly lower silica content (ranging from 0.1 wt.% for a 10% drug load to 0.6 wt.% for a 60% drug load), indicating that uniform coating with less silica effectively improved tabletability.

This investigation provides valuable insights into the superiority of the SAC-based approach over the traditional 1 wt.% silica addition method. It also highlights the limitations of conventional metrics in capturing improvements in flow properties of dry coated materials and demonstrates the impact of coating processibility on feedability and tablet attributes. The findings support the benefits of the dry coating process in enhancing tablet processibility at an industrially relevant scale.

694b Formulating Amorphous Solid Dispersions into Oral Dosage Forms: A Fast to Market Integrated Approach

Patricia Cruz, Inês Ramos, Clara Sá Couto

A large percentage of new drug candidates in the R&D pipeline exhibit low bioavailability due to limited dissolution in the gastrointestinal tract, which can be improved by formulating them as amorphous solid dispersions (ASD). In an ASD, the poorly water-soluble drug is dispersed within a polymeric matrix, stabilizing the drug molecules in an amorphous state, which is more soluble than the crystalline state. Therefore, the development of oral dosage forms using ASDs has drawn significant attention in the pharmaceutical industry due to its potential for improving solubility of poorly water-soluble drugs, leading to increased dissolution rates and ultimately higher bioavailability. This strategy ensures that patients receive the full therapeutic benefit of the medication

This abstract describes an integrated, lean and material sparing methodology for formulation development of ASD-based oral dosage forms, encompassing (i) selection of ASD formulation, (ii) bridging particle engineering with definition of tablet formulation, (iii) analytical characterization, and (iv) scale-up. A case study will be presented to depict this methodology.

Selection of ASD formulation

The selection of an appropriate stabilizing polymer and drug-to-polymer ratio is crucial to achieve the desired drug dissolution performance and physical stability. This optimization process is tailored to the specific characteristics of each drug aiming at maximizing performance, stability and patient compliance by minimizing table size / number of tablets per day. The physical stability of the ASD is assessed through *in silico* screening by estimating the drug-polymer miscibility using the Flory-Huggins and machine learning models, the likelihood of phase separation during droplet drying, and the glass transition temperature. Based on API relevant features, the most suitable processing technology, such as hot melt extrusion (HME), spray drying, or co-precipitation, is selected to manufacture the ASD prototypes as intermediate powders that are representative. Finally, the lead prototypes can be chosen based on biorelevant dissolution performance and solid-state characterization, already considering accelerated stability data, to ensure the quality and performance of the final product.

Bridging particle engineering with the definition of tablet formulation

A robust formulation database is essential for risk-based decision-making during the formulation development process. Leveraging prior knowledge on the processing of spray dried intermediates as tablets/capsules and a consolidated empirical database, suitable excipients and their respective concentration ranges can be identified.

For the final tablet formulation, experimental designs, such as factorial designs, are employed to systematically evaluate the impact of ASD attributes, namely particle size distribution and bulk density, on tablet formulation and process development. Typically, not more than four prototypes are necessary to facilitate comprehensive formulation evaluation with minimal material consumption. Moreover, by formulating with representative materials of larger scales, scale-up challenges may be anticipated and mitigated through particle engineering coupled with downstream process development, aiming at low-risk scalability of the developed formulations which is particularly relevant for First-in-Human supply where typically API availability is limited and cost-effective development is of utmost importance.

Analytical characterization

Throughout development, fit for purpose analytical characterization methods are employed to assess the physical and chemical properties of ASD-based oral dosage forms. Imaging methods, such as Raman imaging and micro-CT, offer insights into the distribution of excipients, solid-state characterization, and tablet surface characteristics. Physical attributes, including powder flow, rheology, and solid fraction measurements, provide valuable information for optimizing formulation performance and manufacturability. Furthermore, biorelevant dissolution studies aid in process and formulation optimization by simulating physiological conditions and predicting *in vivo* performance. This holistic approach to analytical characterization ensures the development of robust formulations with enhanced bioavailability towards consistent therapeutic efficacy.

Scale-up

Computational tools play a pivotal role in accelerating formulation development and optimization processes. Roller compaction modeling offers insights into the process parameters required for direct conversion from slugging to roller compaction, while also highlighting potential challenges such as the loss of compactability using unified compaction curves.

Additionally, compaction simulation studies allow to anticipate and address scale-up issues by simulating large-scale behavior using minimal material. By accurately predicting the behavior of formulations under industrial compaction conditions, potential challenges can be identified and addressed proactively, thereby streamlining the scale-up process.

This presentation intends to show the numerous benefits of an integrated methodology coupled with data-driven tools for pharmaceutical development. Reduction in material consumption and shortened development times are achieved through the optimization of formulation parameters and the utilization of predictive modeling techniques towards reducing the risk for clinical supply. Consequently, ASD-based oral dosage forms are enabled for commercial viability with minimal effort, thereby addressing critical challenges in drug development and enhancing patient access to efficacious therapies.

694c Particle Engineering Approaches to Enhance Dissolution Rate and Ensure Redispersibility

Akshay Korde, Kartik Kamat, Anura Indulkar, Mengqi Yu, Manish Kelkar, Aaron L Featherston, Troy Reynolds, Nandkishor (Nandu) Nere

With increasing molecular weight and hydrophobicity of small molecule active pharmaceutical ingredients (APIs), achieving a high bioavailability has become more challenging. These Biopharmaceutics Classification System (BCS) class II drugs have high permeability but low solubility in bio-relevant media and require enabling formulations such as amorphous solid dispersions (ASD) to increase their solubility. However, screening for the right polymer and excipients combination for achieving a stable ASD can require extensive efforts. Moreover, these ASDs usually have a low drug load and consequently result in a higher pill burden.

Crystalline formulations with micronized API that can lead to a higher dissolution rate, present a viable alternative since most APIs are isolated as a crystalline solid. Particle size control in the final API crystallization then becomes critical and can be achieved via either a top-down (dry or wet milling of larger crystals) or bottom-up (crystallizing small particles) approach. Irrespective of the approach, very small particles tend to agglomerate during drying, and the agglomerated crystals do not redisperse to their original small size, resulting in a lower dissolution rate.In this work, both top-down and bottom-up approaches were explored for the micronization of a poorly water-soluble API and their dissolution performance was compared. The challenges associated with preserving the nanocrystals obtained from the bottom-up approach were addressed by employing different drying strategies, including the use of excipients to reduce API-API crystal interactions. This work presents a reliable strategy that can be leveraged to obtain nanocrystals that can redispersed to their original size in aqueous media.

Disclosure:

Akshay Korde, Kartik Kamat, Anura Indulkar, Miki Yu, Manish Kelkar, Aaron Featherston, Troy Reynolds and Nandkishor Nere are employees of AbbVie. All authors may own AbbVie stock. AbbVie sponsored and funded the study; contributed to the design; participated in the collection, analysis, and interpretation of data, and in writing, reviewing, and approval of the final publication.

694d Tailoring Powder Properties of Drug Substance through Polymorphic Transformation

Elizabeth Horstman, Chiajen Lai, Darren Bykowski, Vijay Dhand

An active pharmaceutical ingredient (API) with a propensity for forming large crystals requires stringent particle size control. Jet-milling, an intuitive top-down approach, was attempted first to meet the particle size target. The micronization process resulted in material that achieved the particle size target but powder properties such as cohesiveness and powder flow can be further optimized. There is a development opportunity which includes enhancing powder properties to enable more robust product-collecting operations and formulation process scalability. In this work, a bottom-up approach using polymorphic transformations is used to achieve targeted particle size and powder properties.

The API has a diverse form landscape of >10 polymorphs and several salts. Form III is the most thermodynamically stable form, and the target form for our process. Two routes were developed to reach Form III with tailored particle properties – one through intermediate Form I and the other through intermediate Form IX. Although Form I and Form IX are metastable polymorphs, they can be reproducibly made through breaking the sodium salt of the API. The final conversions to Form III were effective in producing particles within the size control limit. When high-shear wet milling (HSWM) was integrated into the process, it further enhanced the kinetics of the process. The benefits of this approach include inherently small and uniform crystals, well-defined shape, good flowability and minimal cohesiveness. The investigation demonstrated a bottom-up process strategy used to achieve the desired physical characteristics beyond particle size target. We believe the use of polymorphic transformation to achieve desired properties has broader applications to other pharmaceutical compounds with powder property challenges. The options created by polymorph transformation represent alternative opportunities to ensure formulation robustness and product performance.

694e The pH-Dependent Amorphous Precipitation Behavior of a Small Molecule API

Hursh Sureka, Emily Cunningham, Kathleen Lauser, Bahar Inankur, Maryll Geherty

The amorphous precipitation of APIs is less understood than crystallization, but as small molecule APIs become more complex, precipitation may become more commonplace for their isolation. This presentation will focus on a case study of a small molecule API that was isolated via precipitation and exhibited unpredictable filtration behavior during the isolation, which also resulted in greater entrained solvent and slower drying. Literature suggests a variety of root causes including temperature and mixing. These were investigated, but the strongest contributor was found to be the pH of the slurry. The API was precipitated into water and has a pKa close to neutral but only has trace solubility, so slight basification of the slurry could fully deprotonate the API. However, previous studies had demonstrated both a potential for epimerization and hydrolysis impurity formation at higher pH, so a systematic study of the impact of pH on both filtration behavior and impurity formation was completed giving a wide range of pH where filtration was improved with minimal impact on product quality and process scalability. Multiple bases used in the process were investigated, and a trace kicker of strong base (0.001-0.002 molar equivalents of KOH), was found to be sufficient to decrease filtration time by up to 2 orders of magnitude, while minimizing impurity formation. This process was successfully demonstrated at 10+ kg scale in multiple batches both for this molecule and another API with similar structure and was found to be robust to long process hold times.

694f Inhalation Powder Development: How to Find a Compromise between Deep Lung Deposition and the Flowability Required for Capsule Filling

Gresse Eva, Lechanteur Anna, Geoffroy Lumay, Evrard Brigitte

Pulmonary drug administration has long been used for local or systemic treatment due to several advantages. Dry powder inhalers emerge as the most promising due to efficiency, ecologic, and drug stability concerns. We developed "engineered-inhalation powders" using the spray-drying technique. Particle morphology, from spherical to deflated shapes, was characterized by the number and the depth of dimples measured from SEM images. We define a new characteristic deflation ratio ξ as the product between the number of dimples and their depth. We showed first that the morphology of inhalable powder can be finely tuned by spray-drying technique when excipients and process parrameter varied. Secondly, we developed stable inhalation powders that simultaneously induced high fine particle fractions (>50%) for two drugs (budesonide and formoterol) due to their deflated surface.

However, such inhalation powders introduces challenges related to cohesive forces playing an important role with small particle size. The second part of our study aims to find an easy industrial method for improving powder flow properties to facilitate the handling of these powders, especially in the context of capsules or reservoir filling. To achieve this, we focus on adhesive blends of micronized spray-dried powder, which includes budesonide and formoterol, with coarse lactose excipient. It is intended to develop a homogeneous blend without overly strong interparticular interactions that could potentially compromise the final pulmonary deposition of micronized powders.

Finally, we will show new methods based of soft-pelletization allowing to improve powder flow without the addition of lactose excipients.

References:

Inhalation powder development without carrier: How to engineer ultra-flying microparticles?, A. Lechanteur, E. Gresse, L. Orozco, E. Plougonven, A. Léonard, N. Vandewalle, G. Lumay & B. Evrard, Eur. J. of Pharm. and Biopharm. **191**, 26 (2023)

Engineered-inhaled particles: Influence of carbohydrates excipients nature on powder properties and behavior, A. Lechanteur , E. Plougonven, L. Orozco, G. Lumay, N. Vandewalle, A. Léonard, & B. Evrard, Int. J. Pharm. **613**, 121319 (2022)

Enhancement of inhaled micronized powder flow properties for accurate capsules filling, E. Gresse, J. Rousseau, M. Akdim, A. du Bois, A. Lechanteur & B. Evrard, Powder Technology **437**, 119576 (2024)

694g Wet Granulation Using a Resonant Acoustic Mixer

Emad Renfroe, Blair Brettmann

Control of powder properties is crucial for industrial processes across the food, pharmaceutical, agriculture and mineral processing industries and granulation is an important tool for providing agglomerated particles with controllable properties. However, existing granulation processes are not readily integrated with other processing steps and are not appropriate for some types of materials. Adding resonant acoustic-based granulation to the toolkit has the potential to widen the achievable parameter space, and, importantly, integrate granulation into chemistry and blending operations that are already being performed on the RAM platform, resulting in process intensification. Here we demonstrate formation of granules with particle sizes of ~1-3 mm in the LabRAM II and examine the formation mechanisms in the context of common wet granulation processes. The RAM granulation process followed here involves first forming a large "doughball" agglomerate and then driving its breakup by evaporating the solvent while impacting the doughball against the container walls. We show that this process is similar to the destructive nucleation model for high shear wet granulation with the solvent evaporation in our case leading to the decrease in the liquid saturation of the doughball and a corresponding decrease in its tensile strength and the acceleration in the RAM establishing the impact pressure when the doughball contacts the walls. This work provides a foundation for granulation process design with a resonant acoustic mixer and through its link to existing granulation mechanisms provides a path to a deeper understanding of the process.

610a Development and Utilization of a Flow Process to Cascade Unstable Intermediates to Efficiently Produce High Quality Product

Heather Boucher, Joseph Imbrogno, Jadid E. Samad, Steven M. Guinness, Michele Buetti-Weekly

In recent years, there has been a shift in API manufacturing from traditional batch methods to continuous operation. Operating continuously affords several benefits over batch, including cycle-time reduction, handling of unstable intermediates and products, enhanced analytical data, and ability to operate at increased temperature and pressure conditions not safely feasible in large batch operations. An end-to-end process, including reaction and work-up steps, is desirable to capitalize on all of these advantages. A two-step telescoped flow process directly coupled with two continuous extraction and separation platforms was developed and optimized.

The two-step chemical transformation first includes a reaction in a packed bed reactor (PBR), followed by a coupling reaction within a plug flow reactor (PFR). The former reaction was optimized by generating a kinetic model for both the reaction and the catalyst deactivation rate. The latter reaction was optimized by continuous DoE studies exploring residence time, reactor temperature, and reagent equivalents since the residence time was very short. The outlet of the PFR was fed directly into the first continuous extraction and separation unit to perform the first extraction. The aqueous outlet was sent to waste and the organic solution containing the API was first collected into a small receiving vessel, followed by being pumped continuously into the second extraction unit where it was washed again. The aqueous stream was again discarded and the organic stream containing the API was collected in a receiving vessel followed by final purification steps in batch. These continuous work-up steps were optimized based on solvent volumes, salt wash concentration, and Robatel motor speeds.

This process benefitted from continuous operation in several ways. In addition to cycle time reduction and quickly moving chemically unstable intermediates into the next reaction or processing step, it also allowed for a greatly increased safety profile across both reaction steps. The PBR utilized much less hydrogen and catalyst inside the reactor at any given time and the small size of the PFR allowed for enhanced heat and mass transfer to efficiently remove the exotherm from the reaction. The unstable nature of the final product solution was also mitigated since the process stream flowed continuously into the first two extractions within one minute. Post all continuous operations, the API in solution remained stable for two weeks, which permitted ample time for bulk collection of material to be isolated in batch.

610b Advanced Modular Continuous Manufacturing of Drug Substance: A Novel Digital Twin and Its Applications for Dynamic Optimization, and Control

Ravendra Singh

Advanced continuous manufacturing (CM) is emerging as a preferred platform to produce active pharmaceutical ingredients (APIs). The future continuous pharmaceutical manufacturing is desired to be based on modular plug and play concept. In this approach, mobile modules consisting of a group of unit operations have been developed that can be easily reconfigured to assemble a scalable pharmaceutical manufacturing process for different products. It is also suitable for on-demand manufacturing such as in the case of pandemic. However, because of different levels of complexity, the development and adaptation of flexible and modular CM plant is still an open area of research. Therefore, a systematic framework of methods and tools are needed to develop, adapt and evaluate the flexible modular CM plant for API manufacturing with reduced time and resources. In particular, a digital twin is critical for quick design, adaptation, optimization, scale up, and control of continuous APIs manufacturing process, allowing for virtual modeling of operations without extensive process development.

In this work, a digital twin for continuous APIs manufacturing process using modular components has been developed and its applications have been demonstrated for dynamic optimization and control system design. The process flowsheet model is the heart of 'digital twin' and consists of mathematical representation of three modules with the options of adding more as needed. The first module is for feeding and dispensing. It consists of a refill unit, feed tank, pump, mixing, and preheater. The second module is for performing the chemical reactions required to produce the target APIs. This module consists of tubular reactors placed inside a heating and cooling jacket. The third module is focused on separation of API from impurities. A continuous chromatographic model is currently used for separation purposes that can be easily replaced with any other type of separation techniques such as continuous crystallization. Currently, the developed model library consists of more than 32 units across 10 types that has been used to generate the integrated flowsheet model. The model has been validated using experimental data.

The applications of the digital twin have been demonstrated for scenario analysis, digital optimization as well as design and development of suitable control architecture for a continuous APIs manufacturing process. The dynamic optimization provides the optimum operating conditions for the manufacturing process that improves the critical quality attributes (CQAs). A suitable control architecture then has been developed using digital twin to track the optimum set points for the key control variables. The developed control architecture is also modular in nature and can be easily adapted for different manufacturing processes of APIs that may have different material and information flows, dead time variations, and tuning parameters. The proposed digital twin can save the time and resources needed for continuous APIs manufacturing and improve the product quality significantly.

The objective of this presentation is two-fold; first to highlight the developed unit operation model library and integrated flowsheet model and then demonstrate its application for dynamic optimization and design of control architecture of continuous API manufacturing processes.

References

1. Singh, R. (2023). Advanced continuous manufacturing of drug. Pharmaceutical Technology.

Trends in Manufacturing, 20-26, May 2023. https://www.e-digitaleditions.com/i/1499609-pharmaceutical-technology-may-2023/19?
2. Singh, R. (2024). Digitalization of a novel advanced modular continuous pharmaceutical drug substance manufacturing process. Pharma Focus Europe. Issue 3, 56-64.

Acknowledgements

This work is supported by the US Food and Drug Administration (FDA) under contract number 75F40121C00106.

610c Biaryl Phosphorinane Ligands Flow Chemistry Scale-up

Simona Slater

A series of Biaryl Phosphorinane ligands was synthesized by AbbVie to aid catalysis in a variety of small molecule active pharmaceutical ingredient (API) intermediate synthesis. The first step of the process to produce the ligands was manufactured in the AbbVie API Pilot Plant via flow chemistry. The flow chemistry process used a combination of plug flow reactors (PFRs) and continuously stirred reactors (CSTRs) to facilitate three consecutive reaction steps. Process analytical technology (PAT), including Fourier Transform Infrared Spectroscopy (FTIR) and continuous High Performance Liquid Chromatography (HPLC), was used to monitor reaction progress. The reaction was manufactured in flow as a safety consideration due to the large volume of n-butyllithium required to produce the desired quantity of product. Novel techniques were used to optimize and derisk the process at the lab scale before scaling up to the pilot plant. Several challenges were encountered at scale, including unexpected precipitation, large reaction autothermal reforming (ATR), and unique equipment requirements, which were addressed with process modifications throughout the process design and manufacturing campaign.

610d Synthesis of a Spiroketone API Intermediate Featuring a Green and Sustainable Telescoped Flow Process

Shuoxun Wang, Frederic Buono

The fast-growing demand for efficient and robust synthesis approaches of new chemical entities for active pharmaceutical ingredients (APIs) and their intermediates has seen a rapid development in continuous processing. In a recent drug development endeavor at Boehringer Ingelheim, our focus was on synthesizing a chiral spiroketone as a pivotal building block. We explored both batch and continuous procedures and successfully developed and implemented a telescoped flow process for spiroketone synthesis. This innovative approach offers substantial benefits including energy and cost savings, enhanced throughput with adaptable small-scale devices, real-time reaction monitoring via process analytical technology (PAT), elimination of laborious starting material purification and intermediate separation, reduced Process Mass Intensity (PMI), and streamlined unit operations. Moreover, by transitioning to larger reactors, this process can be readily scaled up to meet manufacturing demands, ensuring excellent transferability to large-scale production. Notably, the developed telescoped flow process reduces catalyst and solvent usage compared to the original batch procedure, resulting in a remarkable 70% reduction in catalyst cost and a 60% decrease in PMI.

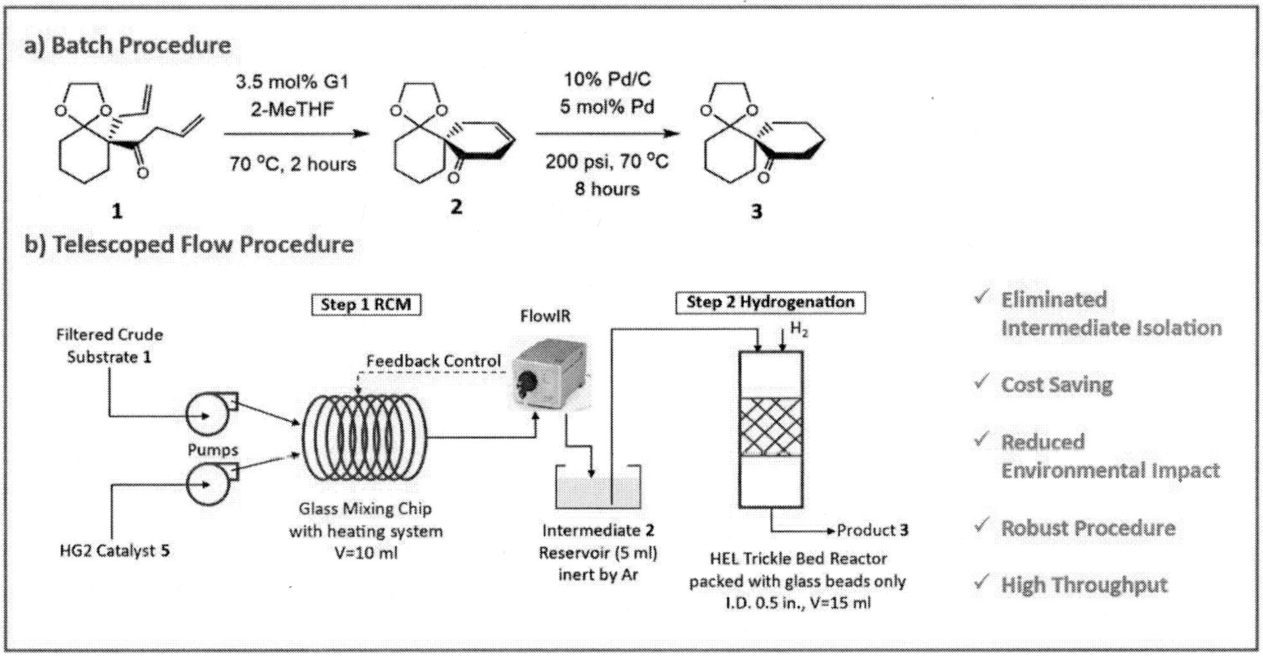

610e Development and Optimization of a Continuous API Manufacturing Process

Katharina Grohowalski, Cameron Armstrong

Continuous manufacturing is often leveraged in the production of active pharmaceutical ingredients (API) and intermediates to enable process intensification, cycle-time reduction, or hazard mitigation. The Flexible API Supply Technology (FAST) group is an engineering focused team that develops continuous manufacturing technology platforms in parallel with assessing the viability of and carrying out continuous processes for portfolio projects in Pfizer's pipeline alongside our process chemistry partners. We present the development of a continuous process along with the application of various automated platforms for reaction optimization in flow applying the use of dynamic/transient flow strategies for ramping residence times and reagent equivalents.

610f Safer Process Design through Continuous Processing – Enantioselective Di-Rhodium Catalyzed Cyclopropanation through a Reactive Aryldiazoacetate Intermediate

Eric Sacia, Stephen Lathrop, Laurie Mlinar, Onkar Manjrekar, Kaid Harper, Steve Richter

Adoption of continuous processes in the pharmaceutical sector remains sporadic despite wide adoption of such processes throughout the petroleum, commodity, and specialty chemicals industry; however, the unique advantages of flow processes to provide dramatically improved heat and mass transfer, to enable differentiated reaction chemistries, and to unlock routes to safer scale-up combine to offer compelling advantages for employing such production strategies. In this presentation, we will discuss the development of a continuous-flow process for the safe formation of a reactive aryldiazoacetate intermediate, extraction of reaction byproducts, drying of the aryldiazoacetate, and its direct use in a fed-batch dirhodium-catalyzed enantioselective cyclopropanation reaction, particularly through the lens of process design and safety.

The process was developed with safety as a primary consideration. In order to enable this scale-up, first the thermal onset temperatures of both the arylsulfonyl hydrazone starting materials and reaction solutions were measured. An appropriate arylsulfonyl hydrazone starting material and organic soluble base was screened to facilitate a Bamford−Stevens diazo-generating flow process at temperatures well below the thermal onset temperature. Reaction conditions were selected that could still facilitate rapid kinetics to minimize aryldiazoacetate accumulation while avoiding conditions prone to thermal runaway. To prevent deactivation of the catalyst in the subsequent cyclopropanation reaction, the Bamford−Stevens reaction byproducts must be efficiently removed. This was achieved via a continuous aqueous extraction using an in-line liquid−liquid membrane separator. Due to the negative impacts of water on the subsequent cyclopropanation, water levels below 100 ppm in the final aryldiazoacetate solution were maintained by adsorption. Finally, the kinetics of the cyclopropanation reaction were then measured at small scale in an Omnical SuperCRC calorimeter and used to predict potential accumulation of the diazoacetate intermediate during fed-batch operation. Performance vs. this model was assessed using Process Analytical Technology, specifically IR, which was also used to monitor reactive intermediate accumulation online during operation.

The complete process was successfully executed on a 100 g scale, setting the foundation for the wider application of this and related chemistries on a kilogram scale in the future within AbbVie. The optimized reaction conditions led to a significant safety advantage in conducting the process in flow, reducing the aryldiazoacetate inventory during operation and increasing temperature control and heat transfer. In conclusion, the study demonstrates a safe, scalable, and efficient process for the formation, extraction, and drying of an aryldiazoacetate and its direct use in a sensitive cyclopropanation reaction. The results provide valuable insights for the design of future flow processes involving sensitive diazo compounds and water-sensitive reactions.

610g Scalable on-Demand Production of Tetrafluoroethylene for Fluoroalkylations in Continuous Flow

Jillian Sheeran, Yuan-Qing Fang, Ali Hasan, M. Grace Russell, David Leahy, Ian W. Davies

A two-staged continuous flow reactor was developed for the on-demand production and consumption of tetrafluoroethylene (TFE). Development of this reactor was motivated by the need for a convenient reactor to perform fluoroalkylations using TFE on intermediate-scales (1 mol/hr TFE) and enable access to bulk material. The efficiency of this reactor was demonstrated by performing the dimerization of difluorocarbene from (trifluoromethyl)trimethylsilane (TMSCF3) to form gaseous TFE (stage 1), which is subsequently reacts with 1,3-butadiene in a gas-phase, 2+2 thermal cycloaddition (stage 2) to produce 1,1,2,2-Tetrafluoro-3-vinylcyclobutane in ~60% yield by 19F-NMR. Key features of this development include (1) use of a multistage plug-flow reactor (PFR) to enable access to high temperature (120-140°C) and high pressure (>250 psi) operation conditions to reduce the reaction time of the 2+2 thermal cycloaddition from 8 hrs to 2 hrs, (2) switching to a high-boiling solvent (diglyme) from tetrahydrofuran and acetonitrile to improve isolation and purification of the cycloalkylation product, (3) continuous distillation of the final product using a wiped film evaporator, and (4) incorporation of automation and safety controls to eliminate human exposure to carcinogenic chemicals, such as TFE. These features enabled the optimization and characterization of a simple fluoroalkylation reactor applicable for multikilogram-scale production of the desired product.

422a *Process Automation Development of 3D Printed Micro-Factories*

Entvin Mamo Sr., Steven Ferguson

Additive manufacturing techniques have become increasingly widespread, with a range of applications including food industry to rocket engineering. However, these methods seem not to have been adopted in the chemical industry. Main reasons are that printed parts typically need to handle high pressure and temperature, harsh chemical environments, operate robustly for long hours and reused. Tackling these issues, we propose the 3D printing of a PEEK microfactory. Creating and testing in practice a flexible PEEK systemized framework for continuous chemical processing[1]. An end-to-end printed configuration converting reagents and/or solvents to formulated products. Capable of continuous flow synthesis with most of the operations, as pressure setting, automated. Compact structures that are versatile to redesign and rearrange to adapt to specific process requirements. This printed configuration is designed to perform multiple unit operations in flow, including reaction, liquid-liquid extraction, and separation. Focus is on achieving end-to-end formulation in a single chip. This reaction platform incorporates custom-designed and methodically calibrated components like an air driven back pressure regulator, inline printed mixing patterns and use of capillary membranes. To demonstrate its commercial potential industrial equipment is integrated. A Bronkhorst pressure controller along with Festo solenoid valves is utilized to manipulate nitrogen from a high-pressure gas cylinder. Temperature and pressure sensors provide constant monitoring of the values among distinct domains. Correlation of process information and chemical analysis data provides valuable and interpretable information to efficiently assess manufacturing system performance. All gathered data is coupled with algorithmic optimization routes. Material selected for the project, PEEK has a superior strength-to-weight ratio, along with excellent chemical resistance and high heat capacity. Furthermore, is a fully recyclable material. These facts render PEEK filament an ideal additive material.

[1] Christopher P.Breen, Anirudh M.K.Nambiar, Timothy F.Jamison, KlavsF.Jensen. Ready, Set, Flow! Automated Continuous Synthesis and Optimization. Volume 3, Issue 5, May 2021.

422b Palladium Removal Using Multistage Centrifugal Extractors and Process Modeling for Design Optimization

Guanghui Zhu, Andrew Cashmore

Extraction and phase separation are essential unit operations in synthetic molecule drug substance processes. There has been a lot of focus on improving synthesis and crystallization process units, however, extraction and phase separation processes are usually overlooked despite their major contributions to process cycle time and process mass intensity.

In this work, we evaluated a multistage centrifugal extractor for palladium removal after a Buckwald coupling reaction. After selecting the scavenger and establishing a baseline with batch extraction experiments, the extraction was studied in a 4-stage centrifugal extractor with varying flowrates and rotation speeds. Dynochem extraction model is used to fit the mass transfer coefficient. A first-principal model is then built to optimize flowrate and number of stages for process design.

Our calculations showed that although centrifugal extractor runs under mass-transfer-limited regime, compared to batch extraction units, centrifugal extractor can significantly improve the process throughput per unit space and improve the sustainability factors by reducing the amount of scavengers and solvents.

422c Complete Synthesis for Albuterol Sulfate in Flow: Development of a Continuous Process for SN2 Amination

Kimberly Penzer, Justin Turnage, Kaitlin Kay, Daniel Gregory, Frank Gupton, James K. Ferri

Continuous processes are finding increased adoption in pharmaceutical manufacturing due to advantages provided over traditional batch processing such as reduced footprint, cycle time, and manufacturing costs in addition to improved quality control. We propose a small-scale, modular, continuous manufacturing system to capitalize on these advantages and address critical shortages of lifesaving drugs like Albuterol. This talk highlights the optimization of the first step of the reaction sequence for the end-to-end continuous manufacturing of Albuterol Sulfate. Kinetic parameters were collected through batch experiments and used to develop conversion models for tubular reactor design. The model and the reactor designs were validated in a series of flow experiments.

Batch amination reactions were conducted at a range of temperatures (0-70 °C) and reactant concentrations (C_{Diol} = 10-100 mg/mL) to verify the reaction mechanism, determine the reaction rate constants, and optimize the amination reaction. Quantitative and qualitative analysis of the intermediate product purity was determined by means of LC-UV and LC-MS. Batch reactions were conducted at 20, 40, and 60 °C in MeOH and IPA. Reaction rate constants and activation energies were determined for comparison between theory and experiment in tubular flow reactors. Continuous amination reactions were conducted in a Vapourtec E-series flow reactor for a range of temperatures (30-60 °C), residence times (3.67-60 mins), and inner tubing diameters (1/16-1/8 in.). The overall flow rate of reactants was maintained at a 4:1 molar ratio, and the temperature was maintained constant. Samples were collected at the outlet of the flow reactor and analyzed *via* HPLC to quantify the presence of each major species.

At low temperatures (0 °C), the reaction rate was slow, limiting formation of the major impurity, a dimer. Higher temperatures increased dimer formation and reduced production of the desired product. An optimal temperature range was observed to exist from ~20-40 °C. Conversion of the diol to the amination product at 40 °C was largely consistent at concentrations ranging from 20-100 mg/mL, with reduced conversions observed at dilute concentrations (*i.e.*, $C_{Diol} < 20$ mg/mL). Kinetics studies demonstrate that the amination reaction is a second order reaction. The experimental exiting concentration of the starting material as a function of residence time was in reasonable agreement with reactor design models for first and second order reactions. This suggests that the experimental conditions in flow nearly approximate pseudo-first order reaction in laminar flow.

422d PAT-Enabled Automated Process Control: Application to Otezla Continuous Manufacturing

Anirudh M.K. Nambiar, Hsiao-Wu Hsieh, Nandini Sarkar, Kartik Saigal, Matthew Beaver, Daniel Griffin

Continuous manufacturing (CM) has the potential to realize some of the visions of Industry 4.0 like lights-out manufacturing where processes run for significant periods of time with minimal manual oversight or intervention. However, to get there, advances in monitoring and control must be proposed, approved, and implemented under GMP drug substance manufacturing.

We report the development and lab-scale demonstration of a feedforward control loop that can adapt to input disturbances to maintain optimal operation during the continuous synthesis of Apremilast, the drug substance of the commercial product Otezla®. The closed-loop control strategy applies process analytical technology (PAT)—mid-infrared spectroscopy, in this case—to monitor feed tank solution compositions in real-time. It then adapts to off-setpoint compositions by automatically adjusting pump flow rates using a Python-based controller to maintain optimal equivalency of the reactants going into the plug flow reactor.

This demonstration helps make concrete the potential for PAT-enabled process control to maintain optimal operation in drug substance CM processes. We will also highlight areas where technical progress is still needed to implement closed-loop control in a manufacturing environment.

422e Automated, Data-Rich GMP Manufacture of a Telescoped Process to Produce API at Commercial Scale with Online Analytics

Joseph Imbrogno, Jadid E. Samad, Heather Boucher, Steven M. Guinness, Michele Buetti-Weekly

Pfizer's Flexible API Supply Technologies (FAST) group has spent the past 5 years focused on developing continuous processes, as well as bringing new facilities online to manufacture at scale, for our small molecule portfolio. A modular design approach was taken to de-risk scale-up and utilize the same modular equipment across our global manufacturing facilities. As of early 2024, our internal kilo lab manufacturing facility was approved for GMP use and was utilized to manufacture an active pharmaceutical ingredient (API) very shortly thereafter. This work will describe the development path to scale-up, implementation of new modules, incorporation of on-line and in-line process analytical technologies (PAT), and fully integrated automation through DeltaV to produce an API for our portfolio.

The first process that was transferred for internal kilo lab manufacture consisted of in-line filtration and carbon pre-treatment, two reaction steps, and two continuous extractions and separations. Both reaction steps were continuously monitored via on-line UPLC samples and in-line FTIR, providing a robust data-rich process. All process parameters were controlled and monitored using DeltaV automation software, through which any deviations could be diverted manually, via PAT, or based on process parameters (parametric control).

This process benefitted from being operated continuously for several reasons, some of which include chemically unstable intermediates, short residence times, exothermic reactions, high catalyst utilization, and direct coupling of downstream purifications steps to remove excess reagents that could lead to degradation. To show operational robustness, the campaign was broken into three separate runs to demonstrate multiple start-ups, shutdowns, and steady state periods. Throughout these three runs, 10 kg of isolated crude product were produced, after final purification steps were conducted in batch mode. Additional modules are still being designed and constructed to expand both the depth and breadth of unit operations that can be utilized to produce small molecule intermediates and APIs using continuous processing.

422f Continuous Simultaneous Hydrogenation and Acetylation of 4-Nitrophenol to Paracetamol

Jimin Park, Marta Hatzell, Carsten Sievers, Andreas Bommarius

Paracetamol (i.e. acetaminophen), the active ingredient of Tylenol®, is one of the largest scale active pharmaceutical ingredients (API) in the world. Current methods to produce paracetamol use non-renewable petrochemical feedstocks, hazardous reagents, have poor Green Chemistry metrics, and operate in batch configuration. New processes to paracetamol should be renewable, safe, environmentally friendly, and operate continuously. Lignin derived phenol and its derivatives are a promising source of the aromatic unit structures needed to synthesize many APIs. To that end, we describe methods to continuously synthesize paracetamol from the hydrogenation and acetylation of 4-nitrophenol.

Traditional hydrogenation methods operate using pressurized stirred tank batch reactors. The high temperature, pressure, and solvent loading of these processes introduce safety issues as well as poor Green Chemistry metrics. Continuous hydrogenation has been shown to mitigate the explosive risks of hydrogenation in industrial processes through reducing required hydrogen headspace and has additional advantages of consistent product quality and superior Green Chemistry metrics.

Careful design of the reaction conditions is required to achieve high selectivity, as the key intermediate of the hydrogenation reaction, 4-aminophenol, is highly reactive and undergoes various side reactions. In previous work, we showed that simultaneous hydrogenation and acetylation of 4-aminophenol in one vessel was highly effective for improving selectivity in a fed-batch mechanochemical reactor. Importantly, careful control of the rate of fed batch acetic anhydride addition was vital for maximizing selectivity.

Here, we implement the reaction engineering strategies learned during our design of our fed-batch mechanochemical synthesis in a flow reactor for continuous, selective paracetamol synthesis with high volumetric productivity. To our knowledge, this is one of the first continuous syntheses of paracetamol from the hydrogenation of 4-nitrophenol with elemental hydrogen.

422g Development of Robust Design Space for Integrated Pharmaceutical Processes through Uncertainty Analysis

Kensaku Matsunami, Zoltan K. Nagy

A hybrid modeling strategy, integrating both a mechanistic and a data-driven model, has been acknowledged as a potent tool, particularly in the context of Industry 4.0 [1]. By combining the strengths of both mechanistic and data-driven models, hybrid models effectively mitigate the limitations inherent in each approach. The approach is beneficial, especially for modeling pharmaceutical processes, which involve a lot of complicated phenomena and a large variety of materials and process settings. There have been numerous applications of hybrid models to pharmaceutical unit operations, e.g., dry granulation [2] and wet granulation [3]. While understanding model uncertainty is critical to the development and calibration of a model [4,5] limited studies have been performed on uncertainty analysis of hybrid models. The structure of model parameters is complex in hybrid models, which makes it difficult to identify improvement opportunities.

This study presents a methodology of uncertainty analysis of hybrid modeling in integrated pharmaceutical unit operations. We consider both drug substance and drug product manufacturing processes, and investigate the robust design spaces for the individual unit operations and the integrated manufacturing systems. For drug substances, a reactor-crystallizer process will be considered whereas for drug products a dry granulation system consisting of a roller compaction unit and a mill will be investigated. Hybrid models are developed for the model systems using the open-source simulation platform PharmaPy [6] and uncertainty analysis is performed based on a two-step approach, as presented by Matsunami et al. [7]. Firstly, probability density functions of mechanistic model parameters were computed based on Markov-chain Monte Carlo (MCMC) simulation. Secondly, a Monte Carlo simulation was performed for a data-driven ribbon milling model considering the uncertainty bounds obtained in the first step. As candidates of the data-driven modeling approach, the multi-linear regression (MLR) and Neural Network (NN) models were tested, whereas NN showed higher prediction accuracy in the previous study [2]. By comparing the uncertainty bounds generated in the first and the second steps, the uncertainty contributions of mechanistic and data-driven models could be quantified.

The results of the uncertainty analysis clarified the propagation of model and parameter uncertainties in the hybrid modeling framework. A deeper analysis was possible by checking the results of each step. The first MCMC simulation visualized uncertainty caused during the model calibration procedure as well as the dependencies of the model parameters. Based on the results in the second step, the suitability of data-driven model approaches was able to be judged based on uncertainty ranges as well as prediction accuracy. The contributions of mechanistic and data-driven models on the uncertainty were simulated, which could suggest which part of the hybrid model should be improved first. The results were compared with the case study performed for the hybrid model of twin-screw wet granulation [7].

In conclusion, uncertainty analysis could provide vital information to improve the performance of hybrid modeling. The validity of the presented approach was confirmed by comparing the results of the dry granulation model and the wet granulation model. Further studies would be expected to increase the applicability and robustness of the approach.

[1] J. Sansana, M. N. Joswiak, I. Castillo, Z. Wang, R. Rendall, L. H. Chiang, M. S. Reis. Recent trends on hybrid modeling for Industry 4.0, Computers & Chemical Engineering, 151, 107365 (2021).

[2] Y. Huang, R. B Lagare, P. Bailey, D. Sixon, M. Gonzalez, Z. K. Nagy, G. V Reklaitis. Hybrid model development and nonlinear model predictive control implementation for continuous dry granulation process, Computers & Chemical Engineering, 183, 108586 (2024).

[3] A. A. Barrera Jiménez, K. Matsunami, D. Van Hauwermeiren, M. Peeters, F. Stauffer, E. dos Santos Schultz, A. Kumar, T. De Beer, I. Nopens. Linking material properties to 1D-PBM parameters towards a generic model for twin-screw wet granulation, Chemical Engineering Research and Design, 193, 713–724 (2023).

[4] D. Casas-Orozco, D. Laky, J. Mackey, G.V. Reklaitis, Z.K. Nagy, Reaction kinetics determination and uncertainty analysis for the synthesis of the cancer drug lomustine, Chem. Eng. Sci., 275, 118591, 2023.

[5] Y. Barhate, H. Kilari, W.L. Wu, Z.K. Nagy, Population balance model enabled digital design and uncertainty analysis framework for continuous crystallization of pharmaceuticals using an automated platform with full recycle and minimal material use, Chem. Eng. Sci, 287, 119688, 2024.

[6] D. Casas-Orozco, D. Laky, V. Wang, M. Abdi, X. Feng, E. Wood, G.V. Reklaitis, Z.K. Nagy, Techno-economic analysis of dynamic, end-to-end optimal pharmaceutical campaign manufacturing using PharmaPy, AICHE J., 69 (9), e18142, 2023.

[7] K. Matsunami, A. A. Barrera Jiménez, T. De Beer, I. Nopens. Uncertainty analysis of hybrid modeling: A case study of partial least squares and population balance model in continuous twin-screw wet granulation, 2023 AIChE Annual Meeting, 327e (2023).

282a Advances in Population Balance Modeling and Simulation

Richard Braatz

Population balance modeling has been a key tool in the understanding, design, and control of a wide variety of processes including crystallization, precipitation, aerosol formation, and bioreactors. While the overall framework was developed more than a half century ago, innovations in numerical algorithms have continued to occur and the number of new applications has continued to grow. This talk will describe some of these developments, both in terms of new application areas such as in gene therapy manufacturing and in terms of numerical algorithms which, in some cases, have orders of magnitude reductions in calculation time. The latter have enabled population balance models to be incorporated into optimal control algorithms that are computable in real time on cheap control hardware.

282b Professor Ramkrishna's Influence on Mathematical Modeling and Analysis of Pharmaceutical Applications

Gintaras V. Reklaitis, Zoltan Nagy, Francesco Rossi

Over his highly productive career Professor Ramkrishna has pursued pioneering mathematical modeling and analysis in a number of areas of importance to the pharmaceutical industry. This body of innovative research, which has motivated and stimulated numerous research efforts across our discipline, has included multidimensional population balance modeling, cybernetic modeling of bioreaction networks, stochastic and probabilistic modeling, and pharmacokinetic / pharmacodynamic modeling. In this presentation we will outline several applications of these modeling concepts that have drawn inspiration from his research, all of which have been carried out by the Purdue faculty team focused on pharmaceutical engineering developments.

282c ^{13}C Flux Analysis in Metabolism Research: From Cells to Multi-Tissue *In Vivo* Models

Jamey Young

^{13}C metabolic flux analysis (MFA) is the gold standard approach for quantifying rates of biochemical reactions in living cells and tissues. It has been widely applied to assess the metabolic phenotypes of industrial host organisms, but it is now being increasingly used to investigate metabolic disease mechanisms both in cellular and *in vivo* models. Over the past 15 years, my lab has established novel ^{13}C MFA tools and approaches that enable us to probe entirely new aspects of metabolism previously inaccessible to measurement. In particular, we have developed publicly available software packages PIRAMID and INCA that automate the computational workflows of ^{13}C tracer analysis and data modeling. I will discuss ongoing work where my lab has leveraged these tools to investigate the pathogenesis of metabolic dysfunction-associated steatotic liver disease (MASLD) using cultured cells and mouse models. We have recently expanded our *in vivo* ^{13}C MFA platform to enable simultaneous determination of intermediary metabolic fluxes in both glycolytic (e.g., muscle, adipose) and gluconeogenic tissues (e.g., liver) of individual mice based on modeling metabolomics datasets obtained from infusion of stable isotopes. These studies have established ^{13}C flux analysis and the INCA software suite as a comprehensive platform to investigate intermediary metabolism in a wide range of biological contexts.

282d 'Continuing" Tangible Impact through Mathematical Modeling in Bio-Pharmaceutical Industry

Eric Murphy, Pelin Su Bulutoglu, William Ketterhagen, Florian Ries, Nandkishor (Nandu) Nere

Chemical engineers play a vital role in the development of various bio-pharmaceutical processes that achieve consistent product quality, predictable scale-up, and efficient operation. Achieving these goals requires in-depth understanding of both the scientific and engineering aspects of the processes providing ample opportunities for chemical engineers to add value. Due to constrained timelines and intrinsic cost considerations around the drug substance and drug product development of novel, "first in the world" molecules, any attempts to generate process understanding and insights while reducing the expensive and time-consuming experimentation is highly desirable. To this end mathematical modeling can play an important role in conjunction with the chemical engineering principles and conventional wisdom. The presentation will introduce some examples of continued tangible impact made through pragmatic modeling approaches employing computational fluid dynamics (CFD), density functional theory (DFT) and statistical thermodynamics, discrete element method (DEM), population balance modeling (PBM) and applied mathematics. The cases include some of the key process equipment and unit operations that are employed as workhorses in the bio-pharmaceutical industry.

282e Transforming Pharmaceutical Development through Predictive Science and Computational Modelling

Pankaj Doshi

Responding to the challenge of bringing lifechanging treatments and medicines with the utmost quality standards to the patients world-wide, the pharmaceutical industry is undergoing a transformational change to reduce the development timelines. Two major transformations which are driving these changes are the transition from batch to continuous manufacturing technology and the digitalisation of product and process design activities. The digital twin of continuous process plays an important role from early to late stages of process development. It allows extensive analysis of the process while minimising experimentation at scale and progress using less material. This is especially relevant to: (i) create process operating space and obtain process information in early stages with reduced API availability, (ii) test new prototypes and changes in equipment, (iii) de-risk the process and investigate extreme and edge points without risk, (iv) understand the process performance and dynamics in cases where it is difficult or even impossible to measure internal variables experimentally, and (v) execute simulations at any time and independently of material availability. The digital twin is further used to investigate the impact of operating conditions on the critical quality attributes of final drug product. It integrates comprehensive process knowledge generated from early to late stage and simulates a range of processing conditions to design an optimal set of experiments. The process knowledge generated from digital twin and other models are getting integrated with the experimental data in the relevant sections of regulatory filing dossier to further strengthen the overall scientific and technical content.

282f Remarks By Professor D. Ramkrishna, His Friends and Collaborators

Meenesh Singh

Impromptu remarks by Professor D. Ramkrishna, his friends and collaborators.

152a Mechanistic Modeling Strategies for Lipid Nanoparticle Production

Pavan Inguva, Saikat Mukherjee, Pierre Walker, Mona Kanso, Jie Wang, Yanchen Wu, Vico Tenberg, Srimanta Santra, Shalini Singh, Allan Myerson, Richard Braatz

Lipid nanoparticles (LNPs) are a versatile drug delivery platform for a wide range of therapeutic modalities including nucleic acids which have received much attention for their use in the COVID-19 vaccines. Modern manufacturing processes for LNPs for nucleic acids involve rapidly mixing an organic stream containing the lipids with an aqueous stream containing the nucleic acids and are conceptually straightforward. However, detailed understanding of LNP formation and structure is still limited and scale-up can be challenging. Mathematical and computational methods are a promising avenue for deepening scientific understanding of the LNP formation process and facilitating improved process development and control. This presentation describes strategies for the mechanistic modeling of LNP formation, starting with strategies to estimate and predict important physicochemical properties of the various species such as diffusivities and solubilities. Subsequently, a framework for constructing mechanistic models of reactor-scale and particle-scale processes is outlined. Insights gained from the various models are mapped back to product quality attributes and process insights. Lastly, the use of the models to guide development of advanced process control and optimization strategies is discussed.

152c Predicting the Long-Term Stability of Biologics with Short-Term Data

Michael Dillon, Jun Xu, Geetha Thiagarajan, Daniel Skomski, Adam Procopio

Understanding the long-term stability of biologics is crucial to ensure safe, effective, and cost-efficient life-saving therapeutics. Current industry and regulatory practices require arduous real-time data collection over two to three years; thus, any method to reduce this drug development bottleneck while still ensuring product quality would enhance the speed of medicine to patients. We developed a parallel pathway kinetic model, combined with Monte Carlo simulations for the generation of 95% prediction intervals, to predict the long-term (3 years) stability of biotherapeutic critical quality attributes (size variants, charge variants, purity by CE-SDS, and potency) with short term (3-6 month) data from intended, accelerated, and stressed storage conditions. We rigorously validated the model with 18 biotherapeutic drug products, comprised of IgG1 and IgG4 monoclonal antibodies (mAbs), antibody-drug conjugates (ADCs), dual protein coformulations, and a fusion protein, including high concentration (≥ 100 mg/mL) formulations and both liquid and lyophilized presentations. For each drug product, we accurately predicted the long-term trends of multiple quality attributes using just 6 months of data. Further, we demonstrated superior stability prediction via our methods with less real-time data than standard linear regression methods. The robust and repeatable results of this work across an unprecedented suite of 18 biotherapeutic compounds suggest kinetic models with Monte Carlo simulation can predict the long-term stability of biologics with short-term data.

152d Optimizing AAV8 Formulation: Coarse-Grained MD Simulations of Salt and Surfactant Effects on Capsid Aggregation

Leila Sharifi, Willow DiLuzio, Arani Chanda, Bodhisattwa Chaudhuri

Purpose:

Adeno-associated virus (AAV) is a non-enveloped virus with approximately 26 nm diameter, which exhibits great potential as a gene delivery tool. However, its manufacturing has challenges such as lower yields, and moderate purity, attributed to degradation during various stages of production. This degradation occurs potentially via aggregation at low ionic strength and insufficient surfactant concentration. These mechanisms of degradation remain poorly understood, and there is a lack of a systematic approach to formulation development. To address this, a coarse-grained molecular dynamics (CG-MD) simulations approach was utilized to understand the molecular level of capsid-capsid interaction under different types and concentrations of salts and surfactants. The model was validated with experimental data.

Methods:

CG-MD simulations were performed using the GROMACS package with MARTINI force fields. AAV8 (PDB: 6v12) with 2.2kb single-strand DNA was selected as the model viral capsid. Energy minimization used the steepest descent algorithm, followed by a 10 ns equilibration step. Production runs were performed using the Nosé–Hoover thermostat and the Parrinello–Rahman barostat at 1 bar pressure for each 20-fs time step in an isothermal-isobaric ensemble.

Results:

We investigated the effect of different types and concentrations of salts on AAV vector aggregation. Our results showed that NaCl and $MgCl_2$ salts at various concentrations could prevent aggregation to some extent. AAV vector aggregation was influenced by the ionic strength of the solution, and NaCl salts were effective at preventing aggregation when present at a concentration above 0.05 M. Additionally, surfactants such as poloxamer 188 at concentrations ranging from 0.001% to 0.01% (w/v) were tested, and our findings suggest that the type and concentration of surfactants are critical factors that affect AAV vector aggregation.

Conclusion

Our developed computational model, with a microsecond timescale, evaluated the behavior of different salts and surfactants with varying concentrations to understand the factors preventing AAV8 aggregation. This analysis provided insights that would be difficult to obtain from experimental studies. The CG-MD simulations proved powerful in simulating macromolecular complexes and guiding the interpretation, and direction of experiments. Our simulations provided important insights into AAV vector formulation development, which could lead to the development of more efficient and stable gene delivery systems.

152e Predictive Model for Glycosylation in Monoclonal Antibodies (mAbs) Using a Knowledge Graph-Based Approach.

Apoorva Nisal, Frantz Gabeau

Abstract:

Glycosylation is the process by which glycans or carbohydrate-based polymers are attached to another molecule. Glycans are critical for storing energy and system regulation [1]. Glycosylation affects the functional efficacy of monoclonal antibodies (mAbs) are one of the main sources of heterogeneity in mAbs. Glycans can exist on a mAb for many reasons such as inherent expression, cell line, cell culture, engineering, substrate, processing, or posttranslational modifications [2]. Glycosylation is structurally diverse and complex posttranslational modification that can affect the behavior of biological molecules. Diverse glycan structures have shown to have effectson the safety, biological activity, and clearance of a molecule.Thus, characterization of glycan species is a critical component in mAb based development to understand how the different species affect the structure and function of the antibodies [3].

Glycosylation includes different types of glycans such as N-linked, O-linked, C-mannosylation, phosphate-linked or glypiation. In biopharmaceutical product development, glycan analysis is conducted using High-Performance LiquidChromatography studies and further extensive characterization using Mass Spectrometry. A HPLC based glycan study sequence includes multiple samples which are processed using a liquid chromatography column. Different glycan species are eluted at different times also known as retention times. Glycan analysis is a time intensive process as scientists must analyze the chromatography results and manually annotate the different glycans based on the observed results.

Previously, a database was created from all the study sequences performed in the past 3 years. A predictive model was developed based off this database to classify and automatically annotate the observed glycans in a chromatogram. The model utilized data such as peak height, width, retention time, plates, and asymmetry to predict the glycan species in a test sample. However, it was assumed that the model is independent of the type of a molecule.

The current approach evaluates this assumption by appending the glycan database with data from monoclonal antibodies. Particularly, this approach evaluates the effect of molecular attributes such as modality, complementarity-determining region(CDR), heavy chain, light chain sequences, molecular weight, isoelectric point (pI), and extinction coefficient on the glycosylation pattern observed in a sample.

A knowledge graph-based approach is proposed to retain all relevant information regarding the monoclonal antibodies and a modified predictive model is developed to further improve the annotation efficiency and reduce manual effort.

References:

[1] Yang, Dandan, Zijing Zhou, and Lijuan Zhang. 'An overview of fungal glycan-based therapeutics.' Progress in molecular biology and translational science 163 (2019): 135-163.

[2] Del Val, Ioscani Jimenez, Cleo Kontoravdi, and Judit M. Nagy. 'Towards the implementation of quality by design to the production of therapeutic monoclonal antibodies with desired glycosylation

patterns.' Biotechnology progress 26.6 (2010): 1505-1527.

[3] Kaur, Harleen. 'Characterization of glycosylation in monoclonal antibodies and its importance in therapeutic antibody development.' Critical Reviews in Biotechnology 41.2 (2021): 300-315.

152f Role of Model-Based Design of Experiments in Pharmacokinetic/Pharmacodynamic Modeling and the Development of Personalized Medicines – Case of COVID-19

Jiaxu Liu, Xuming Yuan, Brahim Benyahia

The advent of personalized medicine has opened promising avenues for the prevention and treatment of diseases by tailoring drug properties and efficacy to the individual needs of patients. This study explores ways to enable the production of personalized solid dosage forms and optimal doses by leveraging personalised pharmacokinetic/pharmacodynamic (PK/PD) models from clinical data. Currently, PK/PD models, which represent the whole population of patients, are built using clinical data obtained from blood analysis and preset sampling procedures e.g. regular sampling intervals. To build accurate personalized PK/PD models, optimal sampling strategies are critical to maximize the data information content and minimize model prediction uncertainties.

A Model-Based Design of Experiments (MBDoE) is proposed to build predictive and reliable personalized PK/PD based on flexible administration and blood sampling strategies and variable drug doses. The MBDoE can be implemented iteratively to reduce model uncertainty below a certain threshold while maximizing drug efficacy and reducing side effects. Consequently, the proposed method can deliver the optimal doses, administration frequency, and sampling strategy for each individual/patient. Consequently, the method can help reduce the overall costs and time of the clinical trials and improve their planning and reliability. Besides, more personalised, and patient-centric medicines can be identified and produced on demand. The initial model PK/PD parameters can be obtained from early clinical trials which can be refined and improved through MBDoE during later clinical trials (Phase 2 and Phase 3).

To demonstrate and validate the capabilities of the proposed method in the context of personalized dosage formulations, a case study focussed on the treatment of COVID-19 using Dexamethasone (Dex) is considered. Through the implementation of systematic MBDoE methodology which optimizes the dosage formulations and timing for blood sample collection, our study aims to reduce inherent risks by reducing uncertainties and facilitating the development of tailored PK/PD models. These reliable personalized models then inform the optimization of the doses and dosage formulations to meet the individual needs, promising a significant strides towards mitigating side effects while preserving and maximizing therapeutic efficacy.

Our findings suggest that personalized clinical trials, underpinned by rigorous PK/PD modelling and optimization strategies, hold considerable promise for enhancing drug safety and efficacy in the treatment of COVID-19 and potentially other diseases. This approach not only contributes to the evolving landscape of personalized medicine but also holds a strategic endeavour towards more patient-centric healthcare solutions.

Acknowledgements

The authors acknowledge funding from the UK Engineering and Physical Sciences Research Council (EPSRC), for Made Smarter Innovation – Digital Medicines Manufacturing Research Centre (DM2), EP/V062077/1.

39a Unveiling the Potential of ML and Small Data in Pharmaceutical Process Development with Sunthericsml

Daniela Blanco, Ethan Tenison

Pharmaceutical process and product development present significant challenges in terms of cost, efficiency, and sustainability. Harnessing the power of data-driven machine learning (ML) to predict behavioral trends, variable effects, and optimal conditions offers a promising solution to fast-track innovation. However, conventional ML algorithms often require extensive datasets, posing a resource-intensive challenge in experimental data collection.

To address this bottleneck, we introduce an enhanced Bayesian optimization (BO) approach, enabling a reaction-agnostic pathway for designing and implementing intelligent experimental campaigns. SunthericsML is an accessible online ML platform tailored for researchers without coding or ML expertise. The pivotal in-lab validation of this approach demonstrates compelling returns on material and experimental efficiency, as well as performance gains against a competitive baseline.

SunthericsML is a versatile technology that allows numeric, discrete, and mixed-integer optimization problems with up to 20 input parameters. This BO-powered approach allows flexible execution in serial or parallel experimentation. Furthermore, it facilitates bounded-target, multi-objective, and constrained-input optimizations, enabling simultaneous enhancements in cost and material efficiency. The versatility of SunthericsML is exemplified through case studies covering chemocatalytic reactions, biocatalytic cascades, and in-vitro mRNA transcription processes, among others.

The in-lab validation of SunthericsML convincingly demonstrates impressive returns on material efficiency, with up to a 75% reduction in the use of expensive or complex reagents. Experimental efficiency sees notable gains, with a 2-6X reduction in optimization experiments. Moreover, the platform enables a 9-12% increase in previously-optimized yields.

39b Model-Based Design of Experiments for Chromatography Method Development

Konstantinos Katsoulas, Federico Galvanin, Luca Mazzei, Maximilian Besenhard, Eva Sorensen

Liquid chromatography is an integral purification process widely used in the pharmaceutical industry for both small molecules and biological macromolecules, separating high-value components of interest from impurities arising from upstream production. Most active pharmaceutical ingredients (APIs) in the drug discovery phase never reach clinical trials, and only a tiny fraction, approximately 1 in 8,000, ultimately gain FDA approval[1]. This presents a significant challenge for pharmaceutical R&D departments, as they must deal with a large volume of separations for numerous molecules with distinct physicochemical properties and behavior. Consequently, the predictive capabilities of computational models have gained increasing interest for their ability to streamline chromatography method development, particularly in parameter estimation, scale-up, and optimization for preparative applications.

Parameter estimation in chromatography aims to identify either fluid-dynamic parameters or (adsorption) isotherm parameters that best represent the underlying physicochemical phenomena of a separation. Selecting the most appropriate isotherm model and accurately determining its parameters is crucial, since the isotherm model defines the adsorption-desorption behavior of the target molecule, which is the key aspect of the separation. Parameter estimation can be achieved through experimentation (e.g., Frontal Analysis) or via hybrid methods, which combine experimental data with modeling tools (e.g., Inverse Method). Although fully experimental methods can be resource-intensive, they generally achieve high parameter accuracy. Conversely, hybrid methods, though preferred due to reduced resource and time demands, have so far tended to be slightly less accurate. The Inverse Method (IM), a prominent example of the hybrid approach, involves fitting model parameters by matching simulations to experimental datasets. A more sophisticated approach is that of conventional Design of Experiments (DoE), or by relying solely on expert process knowledge. These practices, normally focused on a pre-chosen isotherm model, primarily consider statistical tests for goodness-of-fit, but neglect crucial aspects such as practical identifiability and parameter precision.

In our study, we present a rigorous methodology, which employs Model-Based Design of Experiments (MBDoE)[2,3] combined with statistical tests, to evaluate model structure suitability and parameter precision. The method focuses on selecting the most appropriate isotherm model from a set of candidate models, as well as determining its corresponding parameter values and analysing the statistical significance of those values. This MBDoE approach not only identifies the most appropriate isotherm model, but also yields a higher precision of the parameters with fewer experiments compared to conventional factorial DoE approaches; thus, experimental time and resources can be reduced. The approach also considers the precision of the parameter values and the associated model predictions.

In our case-studies, developed based on *in-silico* experiments, we have explored organic modifier-dependent isotherms applicable to both isocratic and gradient separations. The open-source platform CADET in Python was used for the simulations[4]. The parameter estimation employed the built-in gradient descent algorithm of CADET, while MBDoE was performed using a Bayesian optimization algorithm coupled with a Gaussian process as a surrogate objective function.

We will show how accurate isotherm model parameters can be obtained and how the accuracy of the model predictions can be estimated. The use of this methodology will significantly reduce the time and resources required for chromatography model development. The methodology can also identify how reliable the isotherm model is within the given experimental budget, thus providing a very useful tool for business decisions.

References

1. Bhutani, P., Joshi, G., Raja, N., Bachhav, N., Rajanna, P. K., Bhutani, H., Paul, A. T., & Kumar, R. (2021). *Journal of Medicinal Chemistry*, *64*(5), 2339–2381.
2. Franceschini, G., & Macchietto, S. (2008). *Chemical Engineering Science*, *63*(19), 4846–4872.
3. Waldron, C., Pankajakshan, A., Quaglio, M., Cao, E., Galvanin, F., & Gavriilidis, A. (2019). *Industrial and Engineering Chemistry Research*, *58*(49), 22165–22177.
4. Leweke, S., & von Lieres, E. (2018). *Computers & Chemical Engineering*, *113*, 274–294.

39c Model-Based Design of Experiments (MBDoE) for Drug Product Development

Kanishka Ghosh, Salvador García-Muñoz

Design of experiments (DoE) [1] is an important statistical tool that is used extensively in pharmaceutical drug development for the identification of safe operation range, critical process parameters (CPPs) and their effects on critical quality attributes (CQAs), as part of the quality by design (QbD) approach. By manipulating multiple operating conditions simultaneously – all possible combinations (full factorial design) or a portion of the possible combinations (fractional factorial design) – statistical DoE helps identify important interactions for model development and, subsequently, parameter estimate improvement. However, these DoE approaches mainly rely on polynomial model structures postulated to relate operating conditions to process outcomes for experimental design instead of leveraging pre-existing process knowledge, resulting in additional, often expensive, experimentation aimed at re-learning already known process physics. Model-based design of experiments (MBDoE) [2] is a branch of DoE in which the existing physical understanding of a process is leveraged to identify experimental conditions that help gather broader process understanding for further model refinement or parameter precision improvement through potentially lesser and intentional experimental effort.

This work presents an MBDoE workflow that leverages existing process knowledge and extensive experimental information from past campaigns to guide experimental design for the development of a new medicine. Specifically, the workflow assumes *a priori* knowledge about material and process to product relationship (an existing model) and identifies experimental design candidates that, given an experimental budget, are most important for refining the process parameters to accurately explain current system behavior [3, 4]; thus, preventing the design of expansive experimental campaigns such as full factorial design that may not significantly improve upon existing process understanding. The effectiveness of the workflow is demonstrated through optimal MBDoE case studies focused on parameter precision improvement for drug product unit operations such as roller compaction, tablet compression, and film coating. Additionally, this work demonstrates how the workflow can be used to identify the most important sampling times in dynamic processes to reduce over-sampling without losing information gained; a residence time distribution (RTD) identification exercise is used for illustration. Finally, the approach is augmented to deal with a hybrid process model [5] that combines a physical model with a partial least squares (PLS) model.

References

[1] Fisher, R. A., Genetiker, S., & Généticien, S. (1966). *The design of experiments* (Vol. 21). Edinburgh: Oliver and Boyd.

[2] Franceschini, G., & Macchietto, S. (2008). Model-based design of experiments for parameter precision: State of the art. *Chemical Engineering Science*, *63*(19), 4846-4872.

[3] Kusumo, K. P., Kuriyan, K., García-Muñoz, S., Shah, N., & Chachuat, B. (2021). Continuous-Effort Approach to Model-Based Experimental Designs. In *Computer Aided Chemical Engineering* (Vol. 50, pp. 867-873). Elsevier.

[4] Kusumo, K. P., Kuriyan, K., Vaidyaraman, S., García-Muñoz, S., Shah, N., & Chachuat, B. (2022). Risk mitigation in model-based experiment design: a continuous-effort approach to optimal

campaigns. *Computers & Chemical Engineering, 159*, 107680.

[5] Sen, M., & García Muñoz, S. (2021). Development and implementation of a hybrid scale up model for a batch high shear wet granulation operation. *AIChE Journal*, *67*(5), e17183.

39d Using Experimentally Validated ML Models to Predict Stability and Thawing Rates of Large Scale Frozen Biologics

Venkatraman Nagarajan, Tibo Duran, Tanu Mehta, Bruna Minatovicz, Yangchao Luo, Yi Wang, Bodhisattwa Chaudhuri, Saeed Najafian

Purpose:

Freezing and thawing (F/T) of bulk protein solutions is a common processing step used to maintain stability and quality of biopharmaceutical products during development and production. Although F/T of proteins offers numerous advantages, the stresses from freezing and thawing can result in the loss of protein stability and functionality. Therefore, F/T of proteins need to be better understood to maintain stability and quality of proteins during manufacturing. The overall objective of this research is to use a validated machine learning model to predict the thawing rate and the stability of our model protein, BSA after every F/T cycle. Protein stability is evaluated quantitatively using protein aggregation studies such as dynamic light scattering (DLS) and high-performance liquid chromatography (HPLC). Thawing rates and parameters evaluated from protein aggregation studies are obtained experimentally and are then used in the training of the developed ML model to make predictions.

Methods:

An Artificial Neural Network (ANN) model is used to predict the thawing rates and stability of model protein, BSA, after every F/T cycle. Data driven ML models such as the ANN models, have been chosen as they perform well in the case of non-linear experimental data, and they also have the capability of extrapolating data to overcome the complexity and cost burden of experimental work. The input parameters such as protein solution concentration, surface area to volume ratio, loading distance between bottles while thawing, etc. were used as input features in the developed ANN model and the thawing rate and stability parameters were extracted from the model as predictions. Multiple-input-single-output models were implemented in Matlab using the Levenberg-Marquardt (LM) algorithm. In the experimental side, a protein surrogate solution and a protein solution were used to experimentally determine the thawing rate. Both the surrogate and the protein solutions were frozen for 24 hours in a -80°C freezer and then thawed. Thawing rates (°C/min) were calculated based on temperature differences between the starting and ending points of experiments and the time taken to reach the end points from the starting points.

Results:

The goal of the experiments was to calculate the thawing rates and quantify stability in terms of particle aggregation or monomer recovery for the protein solution. The experimentally obtained values were then used to train the developed AI-ML model to make predictions of protein thawing rate, protein surrogate thawing rate and protein stability as functions of F/T process parameters and material attributes. The AI-ML model performance was evaluated using Mean Relative Error (MRE) between the ground truth values and predicted values.

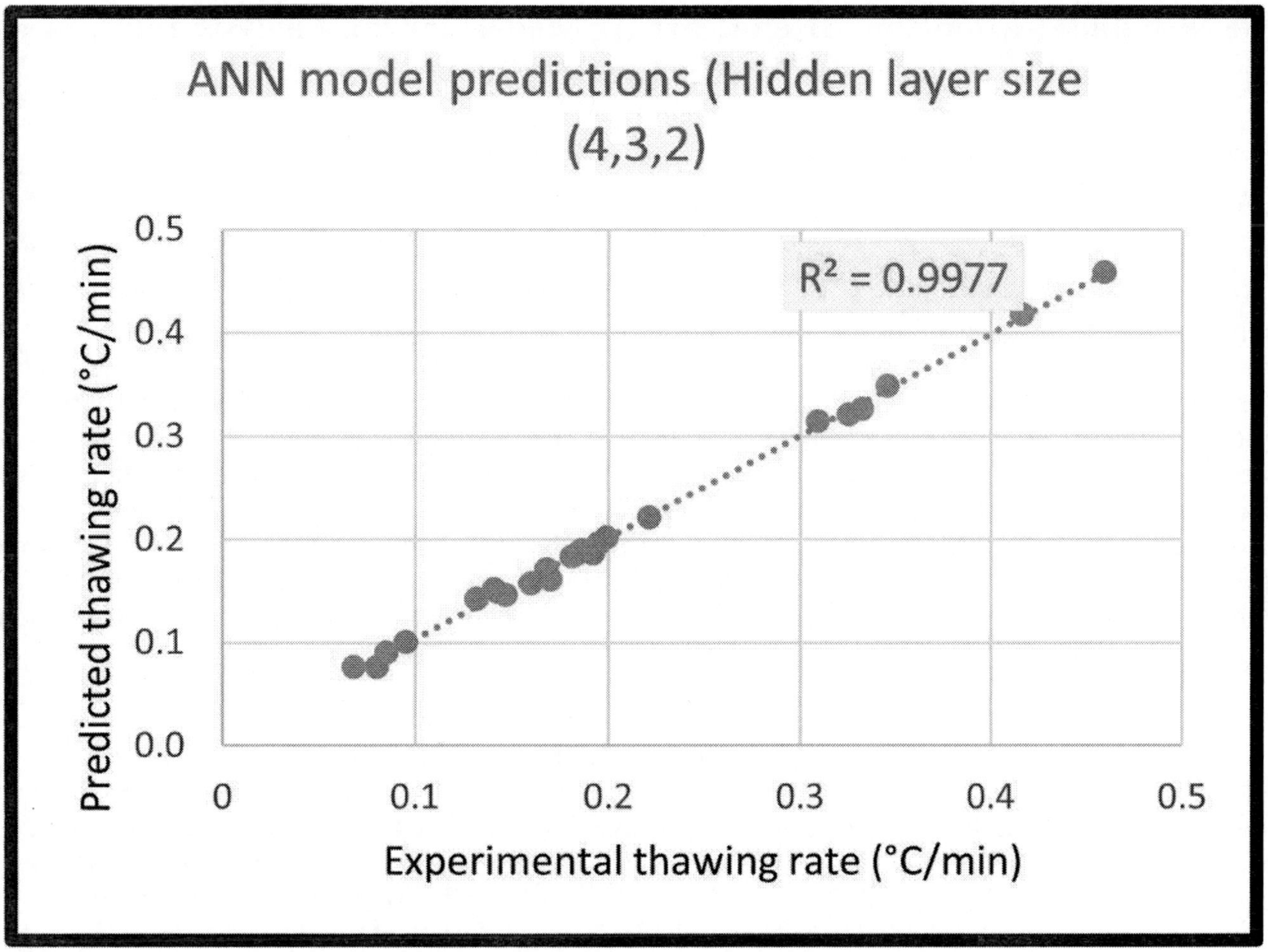
ANN model predictions (Hidden layer size (4,3,2)
R² = 0.9977
Predicted thawing rate (°C/min)
Experimental thawing rate (°C/min)
0.0
0.1
0.2
0.3
0.4
0.5
0
0.1
0.2
0.3
0.4
0.5

39e Machine Learning and Deep Learning Models for the Analysis and Prediction of Pharmaceutical Powder Blend Properties

Joshua Young, Rajesh Dave, Anna Owasit, Srivatsan Jayaraman, Satyam Pandey, Siddharth Tripathi, Mohit Desai

The bulk properties such as flowability of pharmaceutical powder blends and formulations are difficult to predict as the resulting blend flowability is not a linear interpolations of the individual constituent material's flowability. Flowability is crucial for tableting but evaluating it experimentally can be time-consuming and resource intensive. Additionally, when one of the blend components is very cohesive or cohesive, improving flowability through conventional means of blending with a good flowing material can be challenging. To this end, dry coating has emerged as an innovative technique and can be used to tailor cohesion and enhance flowability in a blend, but it further complicates property prediction as the coated blends exhibit remarkably different flowability compared to their uncoated counterparts. To overcome these challenges, machine learning (ML) and deep learning (DL) can be used owing to their ability to learn non-obvious connections within complex chemistry-structure-property relationships. Combining ML/DL with mechanistic approaches may yield better predictions of powder and powder blend properties.

Here, we applied ML/DL methods to (1) predict flowability of powder blends from properties of their constituent API and excipient powders, (2) understand the features most predictive of flowability, and (3) propose new blend formulations with enhanced properties. First, supervised ML models were trained on a dataset of 400 powder blends consisting of varying combinations of 7 APIs and 6 excipients, both with and without dry coating of nano-silica of different types and amounts. Both regression (direct prediction of the flow function coefficient) and classification (prediction of flowability regime such as free flowing, easy flowing, or cohesive) ML models were trained using size and morphology features of the constituent API and excipients, with the classification algorithms strongly outperforming the regression ones. Random forest and XGBoost models, the best performing, predict the flowability regime of the API/excipients blends with up to 85% accuracy. Further feature importance analysis demonstrated that the dry coating amount was most important in determining flowability, followed by the size and morphology features of the constituent API and excipient.

Second, a variational autoencoder (VAE), a type of generative algorithm, was trained on the experimental data. The trained VAE was used to produce 163 new "synthetic" powder blend formulations and predict the flowability of each. Following this, 12 new powder blends predicted by the ML/DL models were experimentally created and their flowability measured. Comparison of their measured flowability regimes to those predicted by the ML/DL models resulted in an accuracy of 83%. In summary, we demonstrated that supervised ML and DL models can perform well at predicting the relative flowability of powder blends based on their components, give insights into powder feature-property relationships, and be used for the proposal of new blends with improved properties. The integration of these computational approaches with mechanistic understanding can enhance process and product development in the pharmaceutical industry.

39f An End-to-End Machine Learning Framework for the Characterisation of Tablet Disintegration

John Armstrong

Tablet disintegration can be described as the combination of four interconnected processes: liquid absorption, swelling, the breaking of interparticle bonds and excipient particle dissolution. All these mechanisms are related to the raw material properties of the tablet as well as its composition and manufacture. Understanding these relationships is important for optimising a tablet's robustness and reliability. The sessile drop experiment is a simple yet powerful tool for testing the liquid absorption and swelling characteristics of a tablet. A single droplet of water is dispensed onto a tablet with a video recorded of the evolution of the droplet as it interacts with the tablet. The following introduces a fully automated method for extracting and analysing data from the sessile drop experiment using a combination of image segmentation, regime detection and model fitting.

The frames of the video constitute the experimental data with image segmentation employed to locate the droplet and tablet construct in each frame for further analysis. This is done by training a U-Net neural network. The U-Net works as a per-pixel classifier, trained by minimizing the linear combination of the binary cross entropy (BCE) and Sørensen-Dice coefficient (SDC) losses. The BCE determines how confident the U-Net is in its prediction of the classifications while the SDC loss is a measure of similarity between the ground-truth and predicted segmentations. This is implemented in PyTorch and trained for 18 epochs until convergence across 4 NVIDIA H100 GPUs.

Once the droplet has been identified in each frame, the surface area and contact angles are extracted from the frame. The surface area is simply a summation of the number of pixels in the droplet multiplied by the physical area of one pixel. The contact angle is found by identifying the edge of the droplet, fitting an ellipse, and finding the gradients of the points where the fitted ellipse crosses the smoothed tablet surface. The angle between the gradient and the tablet surface is then easily calculated. The extracted data are then the surface area at each time , $S(t)$, and the contact angle, $\Phi(t)$.

To characterise the absorption and swelling processes from the surface area, three key timepoints need to be identified: the absorption start time, t_a,i; the absorption end time, t_a,f; and the swelling start time t_s,i. t_a,i is obtained by locating a maximum turning point in as this will be present where the droplet comes into contact with the tablet and signifies the beginning of liquid absorption being the dominant process.

To obtain, t_a,f and t_s,i, the surface area profile is processed by a Sequence to Sequence (Seq2Seq) neural network. The Seq2Seq model works as an autoencoder which classifies each point in the surface area profile as being during the liquid absorption, transition and swelling phases based on information from the previous timepoints. It is trained on all available formulations' data and validated against unseen formulations achieving 96.71% accuracy. Most of the confusion in this model comes from estimating the start of the swelling phase sooner than the manual labelling did but this is not thought to have much effect on the model results. Once all three times are obtained, the surface area can be split into absorption and swelling with mechanistic models fitted to each.

The liquid absorption, $y(t)$, is quantified as the absolute relative change of the surface area at time, t_a,i (equation shown in attached figure) and modelled with a power law function to determine the liquid absorption rate, k, and the absorption linearity, m, of the tablet. The swelling, $\sigma(t)$, is expressed as the change in surface area from the value at time t_s,i. This is modelled using the Schott equation (shown in attached figure) giving three values characterising the swelling – A, the multiplicative reciprocal of the initial swelling rate, σ_∞, the maximum swelling size and t_r, the time delay before the swelling becomes uniform.

This fully automated workflow allows for the rapid, robust analysis of large datasets accelerating understanding of the absorption and swelling processes of tablets. On a modest GPU (NVIDIA RTX A6000), the workflow achieves a 360x speedup over traditional image processing used in previous studies. This takes the analysis from requiring hours of compute per video to full analysis in less than one minute. This has been used to analyse a large dataset of tablets (200) containing the API griseofulvin to compare how the addition of an API changes the disintegration time of tablets. All tablets contained 30% w/w of API, 5% croscarmellose sodium, 1% magnesium stearate and the evenly split between two excipients. For the tablets with dissolution controlled disintegration, the presence of API weakens the correlation between disintegration time and absorption linearity while for wettability controlled formulations the correlation is much the same. Swelling controlled formulations have a more extreme reaction to the API with both liquid absorption parameters changing their correlation signs while still maintaining similar magnitudes. This demonstrates that at high drug loading, there is a great effect on the disintegration performance of dissolution and swelling controlled formulations while wettability controlled formulations behave much the same regardless of API presence.

Another application of this approach is the integration into a fully automated tabletting microfactory. The microfactory consists of a bespoke dosing unit, near-infrared spectral analysis for content uniformity, compaction simulator, hardness tester and bespoke sessile drop apparatus. The products are transferred between the stations via robotic arms. The sessile drop analysis has been optimised to run on an NVIDIA Jetson AGX Orin Development Kit using NVIDIA's TensorRT framework. This implementation works in real-time (max. performance 45FPS) at low power (50W) to analyse sessile drop videos as soon as they are available. The results from the model fitting are combined with modelling approaches of the other instruments into a global Bayesian optimisation so that the setup can work towards achieving a user-defined target e.g. specific tablet porosity or disintegration time.

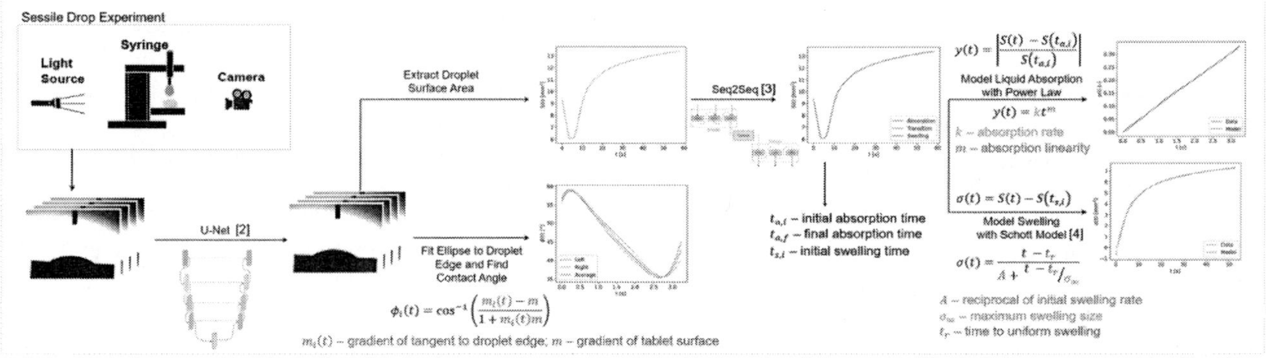

39g Developing Digital Twins for Pharmaceutical Crystallization Processes Using Machine Learning-Based Strategies

Yash Barhate, Neda Nazemifard, Ben Renner, Yihui Yang, Charles D. Papageorgiou, Zoltan K. Nagy

Population-balance-based mechanistic digital twin development has seen increasing adoption within the pharmaceutical industry, facilitating the design of crystallization processes to ensure consistent delivery of crystals with desired attributes such as purity, crystal size distribution (CSD), yield, and polymorphic form. However, accurately developing system-specific population balance models (PBM) presents challenges, including the identification of suitable semi-empirical equations to describe various mechanisms and the estimation of kinetic parameters from experimental data.[1] Moreover, the high computational cost of first principles PBM models hinders their real-time application, whether for updating crystallization kinetics based on new process data or for real-time optimal control under uncertainty. To this end, data-driven digital twin development using state-of-the-art machine learning (ML) and artificial intelligence (AI) methods presents a promising avenue for providing a viable modeling alternative, overcoming the challenges associated with mechanistic digital model development for complex pharmaceutical systems, and in some cases, even surpassing the capabilities of mechanistic digital twins.[2]

While a notable surge in literature demonstrates proof-of-concept studies applying data-driven strategies for modeling and controlling crystallization processes, a significant research gap persists, particularly concerning pharmaceutical APIs. This gap is characterized by limited experimental connections, either through training models with experimental data or validating models against experimental results.[2] Thus, the primary objective of this research is to address this gap by devising systematic frameworks for constructing data-driven digital twins from experimental data and assessing their effectiveness for robust design and control of pharmaceutical crystallization processes.

The framework development was initiated using a model compound from the literature with a needle-shaped morphology described using a complex two-dimensional PBM model.[3] The data-driven digital twin development was conducted in phases mirroring a typical pharmaceutical API development procedure followed in the industry: small-scale Crystalline experiments, lab-scale crystallization experiments, and pilot-scale experiments. For each phase, the training data used for building the data-driven models was constructed using specific experimental information available during each phase. For example, only the operating conditions and crystal quality attribute information (yield and CSD in both length and width directions) were used in the first phase, while the second stage additionally consisted of calibrated process analytical technology (PAT) data from experiments done at lab-scale. The machine learning development workflow included key steps such as data pre-processing, evaluation of different model architectures, optimization of various hyper-parameter settings, and model testing. To address the commonly observed training data-scarcity challenge associated with developing data-driven digital twins from experimental data, state-of-the-art deep learning-based generative modeling techniques were evaluated and used for generating synthetic experimental data from real data. Furthermore, augmentation of real and synthetic data was performed to enhance the training process of neural networks, thereby improving their prediction capabilities. Lastly, the developed data-driven digital twin framework was applied and validated using the experimental data from a real-life API crystallization process.

Acknowledgement:

Funding from Takeda Pharmaceuticals International Co. is gratefully acknowledged.

References:

1. Nagy, Z. K. & Braatz, R. D. Advances and new directions in crystallization control. *Annu. Rev. Chem. Biomol. Eng.* **3**, 55–75 (2012).
2. Xiouras, C., Cameli, F., Quilló, G. L., Kavousanakis, M. E., Vlachos, D. G. & Stefanidis, G. D. Applications of Artificial Intelligence and Machine Learning Algorithms to Crystallization. *Chem. Rev.* **122**, 13006–13042 (2022).
3. Wu, W. L., Chappelow, C., Hanspal, N., Larsen, P. A., Patton, J., Shinkle, A. & Nagy, Z. K. Digital Design of an Agrochemical Crystallization Process via Two-Dimensional Population Balance Modeling. *Org. Process Res. Dev.* **28**, 558 (2023).

219a Addressing Density Functional Theory Modeling Challenges in the Crystals of Highly-Flexible Pharmaceuticals

Gregory Beran

Density functional theory modeling has transformed crystal structure prediction in recent years, and crystal structures of rigid and moderately flexible molecules can be now routinely be predicted accurately. However, significant challenges remain for highly-flexible molecules such as drugs, as will be demonstrated by presenting the contrasting results for molecules XXXI and XXXII from the 7th Blind Test of Crystal Structure Prediction. Whereas moderately flexible molecule XXXI represents a relatively straightforward example for crystal structure prediction, the highly flexible drug molecule XXXII proves extremely challenging. Even obtaining reliable conformational energies for molecule XXXII proves very difficult. General computational strategies for how to handle such challenging cases will be discussed.

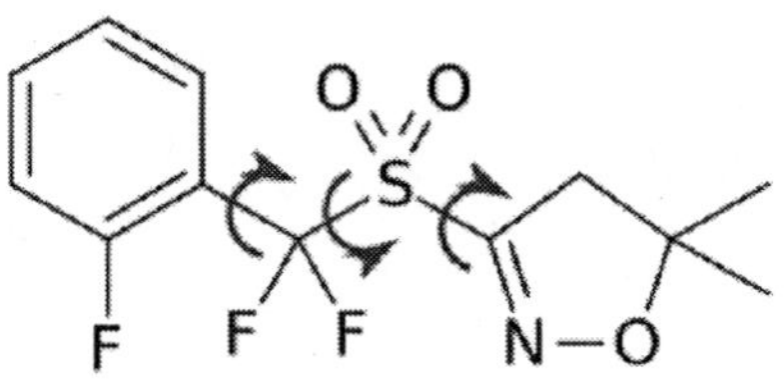

XXXI

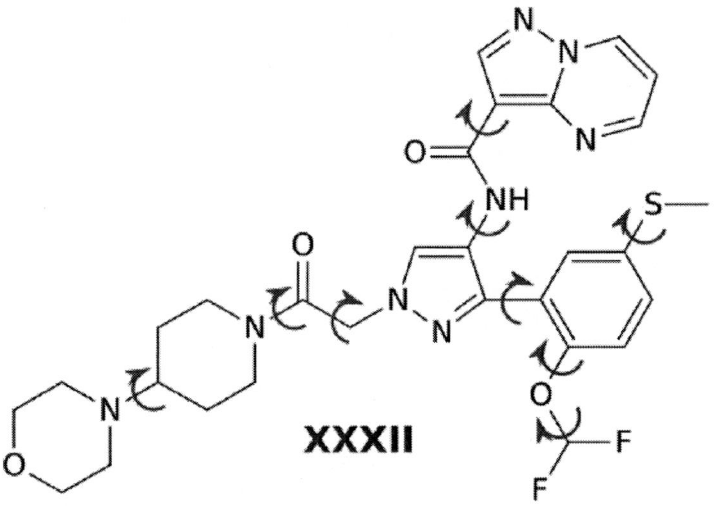

XXXII

219b Characterizing Acid-Base Organic Crystals By Computationally Efficient Periodic System Calculations

Yuriy Abramov, Jian Wang

The definitive characterization of pharmaceutical salt and cocrystal solid forms is crucial from regulatory and IP perspectives. However, the assignment of acid-base configurations along the salt-cocrystal spectrum solely by single-crystal X-ray diffraction (scXRD) experiments may be ambiguous due to the low scattering power of polar hydrogens. Therefore, additional experimental or theoretical studies may be required to confirm the proton location between acid and base. We benchmarked the applicability and limitations of commonly used computationally efficient periodic system calculations for the characterization of acid-base systems against a dataset of 19 crystals with proton positions experimentally assigned using ssNMR or single-crystal neutron diffraction methods. A practical computational workflow to support the characterization of pharmaceutical acid-base crystals in conjunction with routine scXRD measurements was proposed.

219c Symbolic-Numerical Computational Tool for Mechanistic Crystal Growth Models of Molecular Crystals

Neha Padwal, Michael F. Doherty

Controlling crystal properties is vital for product functionality especially in the materials, electronics, and pharmaceutical fields. Crystal growth models enable prediction of crystal properties such as size distribution, shape, morphology, and polymorph, as a function of their growth conditions. On the atomic scale, crystal grows via incorporation of molecules along *steps* mainly at special sites of attachment called *kinks*. This brings about key quantities of *kink density* and *step velocity* in growth models. The major limitations in existing kink density and step velocity models are twofold: 1. Kink density is modeled at equilibrium utilizing the Boltzmann distribution, which ignores spatial correlations between surface sites, and the nonequilibrium nature of crystallization, 2) step velocity models neglect nucleation rates. This underscores the necessity for a new theory to address the existing deficiencies.

We have developed the Simplified Steady-State Framework (SSSF), which addresses the above deficiencies in a systematic approach for Kossel crystal and AB crystals with two growth units in the cell (Z=2).[1,2] The framework consists of a steady-state master equation analysis and is based on accounting for only a small fraction of most probable surface events at the most likely surface sites. The model predictions based on the novel framework have been validated against kinetic Monte Carlo simulations as well as experimental morphology observations for various active pharmaceutical ingredients such as doravirine precursor, celecoxib, ritonavir I, among others.

In this work, a computational model development and implementation tool is proposed to extend the framework to crystals with any number of non-centrosymmetric growth units in the unit cell (Z>2). The tool consists of two separate engines: symbolic and numerical engines. The symbolic engine utilizes graph network theory to represent the surface kinetics in the form of a network, where the surface sites and rate events become nodes and edges within a network, respectively. Such a network representation allows development of the steady-state master equation model equations, which are then supplied to the numerical engine. Within the numeric sections, the equations are then solved simultaneously employing nonlinear equation solvers. The solvers are utilized in conjunction with a parametric continuation scheme, which incrementally increases parameters within growth models from a known mathematical solution at the Kossel crystal point to the final crystal conditions. Such a computational tool will enable application of SSSF to generate accurate growth rate predictions for complex organic crystals with any number of growth units in the unit cell. This modeling approach is expected to play a crucial role in guiding the design, control, and optimization of crystallization processes, whilst yielding crystals with precisely tailored properties.

References:

1. Padwal, N.A. and Doherty, M.F., 2022. Simple Accurate Nonequilibrium Step Velocity Model for Crystal Growth of Symmetric Organic Molecules. Crystal Growth & Design, 22(6), pp.3656-3661.
2. Padwal, N.A. and Doherty, M.F., 2024. Step Velocity Models for Crystal Growth of Organic Molecules: Two Molecules in the Unit Cell. Crystal Growth & Design. (*Accepted for publication*)

219d Computer-Aided Solvent Selection to Control Nucleation and Resulting Product Quality Attributes in API Crystallization

Jacek Zeglinski, Emilia Jakubowska, Tomasz Koczorowski, Marko Ukrainczyk, Brendan Roche, Gary Morris, Barbara Wood, Brian Glennon, Melba Simon

As development timelines for new small molecule therapies become increasingly compressed, attention has turned to the creation of new tools and methodologies to streamline the laboratory process development effort. Computational approaches for the prediction of solubility are now widely used in the design of industrial crystallization processes to aid solvent selection and reduce development timeframes and material needs.[1] However, despite the significant influence of solvents on the nucleation and growth kinetics and thus particle size outcomes, there have been limited examples of computational methods being used to guide solvent selection with consideration of the particle size goals of the process.

A relationship between nucleation difficulty and solvent-solute interaction strength has been predicted computationally and demonstrated experimentally for a range of different organic molecules, strongly suggesting that the desolvation of a solute molecule is one of the key elements governing crystal nucleation kinetics.[2] In contrast to current computational approaches which require both specialized knowledge and supercomputing capabilities, we demonstrate that activity coefficients (which can be easily computed with COSMO-RS software) can be used to estimate the strength of the solvent-solute interaction in solution. Those activity coefficients can serve as *in silico* descriptors, enabling *a priori* prediction of the relative nucleation ease/difficulty of a solute crystallized out of a set of different solvents and the preparation of a ranking list to guide solvent selection.

In this contribution, we demonstrate how this original workflow can be practically applied as part of a high-throughput *in silico* solvent screening to guide the selection of a preferred solvent to facilitate ease of nucleation out of a broad list of solvent possibilities. The insights from this approach can allow crystallization practitioners to assess how different solvents may alter the balance of nucleation and crystal growth occurring in the process to support decision making on what solvents best enable the design goals of the crystallization (*e.g.*, particle size, uniformity, and size distribution).

References

[1] Lovette, M.A.; Albrecht, J.; Ananthula, R.S.; Ricci, F.; Sangodkar, R.; Shah, M.S.; Tomasi, S. *Cryst. Growth Des.* **2022**, 22, 5239–5263.

[2] Khamar, D.; Zeglinski, J.; Mealey, D.; Rasmuson, Å.C. *J. Am. Chem. Soc.* **2014**, 136, 11664–11673.

Acknowledgement

This research was partially supported by the European Union Horizon 2020 MSCA RISE programme, 'ORBIS' grant agreement No. 778051

219e Modelling and Analysis of Roller Compaction Using the Discrete Element Method

Plamen Grigorov, Stefan Pantaleev, Fang Wang, Sabrina Green, Patrick T. Schilling

Roller compaction is commonly adopted in dry granulation processes due to its reliability when dealing with cohesive materials and its suitability for both batch and continuous manufacturing. Despite its widespread use, the mechanics of this process is not fully understood, leading to high reliance on physical trial-and-error in process design and sub-optimal performance in operations. Simulations based on the Discrete Element Method (DEM) can provide unprecedented insight into the mechanics of the roller compaction process and enable rapid and cost-effective virtual trials that inform the relationship between operational parameters and process performance.

In this work we present an efficient DEM modelling methodology for the roller compaction process and a parametric analysis of process mechanics based on DEM simulation data.

The proposed methodology combines meso-scopic modelling with inertial scaling to achieve three orders of magnitude increase in computational efficiency relative to the traditional DEM modelling approach, making the simulation of the full-scale compaction process practical. Material model calibration based on shear cell and uniaxial compression measurements is employed to achieve an excellent prediction of the experimentally observed roller compaction force and mass flow rate at different screw speeds and roller gaps. The model calibration is performed using an automated approach that combines Design of Experiments (DoE), machine learning and genetic algorithm optimization in the interest of efficiency.

A virtual parametric analysis is performed, focusing on the effect of screw speed and roll gap on the specific roll force and the relative density distribution in the compact. The analysis reveals the determining role of the mass flow rate on the roll force as well as a complex interplay between the process parameters and the compact relative density. A spatially non-uniform and temporally variable compact density distribution is observed in all cases. This is in contrast with the results for a compact made by slugging and may partially explain the difference in performance of the two processes. The simulations reveal a complex stress state in the nip region which has a cyclic behavior that can be linked to the motion of the screw and may explain the observed spatial and temporal variance of the relative density distribution in the compact.

219f Predictive Model of Powder Compaction Using Finite Element Analysis

Santhosh Seshadhri, Karthik Chittepu, Geoff Mckee, Sweta Manthena, Pankaj Doshi

The tablet is currently the most widely used dosage form for oral drug delivery. Despite the various advantages of the tablet over other dosage forms, such as superior physical or chemical stability, tablets remain challenging to formulate because their properties depend strongly on the powder composition and details of the compaction process. Powder compaction is a critical process in manufacturing the tablet dosage form. Although this process has been used routinely for over a century, problems related to powder compaction in pharmaceutical formulation development and manufacturing persist. Common problems include tablet failures, such as capping and lamination, high friability, powder sticking to punch surfaces or the die wall, and insufficient mechanical strength to withstand stress in downstream processing.

In this work a workflow was created to de-risk and better understand the behavior of Powder compaction, using the Finite Element Analysis (FEA) by using the density-dependent Drucker-Prager Cap (DPC) model using ANSYS Mechanical. This paper presents a full description of the DPC model for the compaction behavior of microcrystalline cellulose (MCC). In this paper, Finite Element Modelling (FEM) and Design of Experiment (DoE) techniques are adopted to find the impact on optimal shape which has more uniform mechanical properties and less capping and chipping tendency. Response Surface Methodology (RSM) was employed to establish the relationship between the design variables, represented by the geometrical parameters and the friction coefficient, and compaction responses of interest including residual die pressure, the variation of relative density within the tablet, and the relative shear stress of the edge of the tablet.

The results showed that the finite element model will be able to accurately predict the compaction behavior of the MCC powder. Furthermore, the FE predictions of stress and density distributions of the powders during the compaction were used to analyze the failure mechanisms associated with tableting.

219g Development of a Semi-Mechanistic Modeling Framework for Wet Bead Milling of Pharmaceutical Nanosuspensions

Don Clancy, Gulenay Guner, Sayantan Chattoraj, Helen Yao, Connor Faith, Zahra Salahshoor, Kailey Martin, Ecevit Bilgili

This study aimed to develop a practical semi-mechanistic modeling framework to predict particle size evolution during wet bead milling of pharmaceutical nanosuspensions over a wide range of process conditions and milling scales. The model incorporates process parameters, formulation parameters, and equipment-specific parameters such as rotor speed, bead type, bead size, bead loading, active pharmaceutical ingredient (API) mass, temperature, API loading, maximum bead volume, blade diameter, distance between blade and wall, and an efficiency parameter. The characteristic particle size quantiles, i.e., x_{10}, x_{50}, and x_{90}, were transformed to obtain a linear relationship with time, while the general functional form of the apparent breakage rate constant of this relationship was derived based on three models with different complexity levels. Model A, the most complex and general model, was derived directly from microhydrodynamics. Model B is a simpler model based on a power-law function of process parameters. Model C is the simplest model, which is the pre-calibrated version of Model B based on data collected from different mills across scales, formulations, and drug products. Being simple and computationally convenient, Model C is expected to reduce the amount of experimentation needed to develop and optimize the wet bead milling process and streamline scale-up and/or scale-out.

40a New Features for Lipid-Based Drug Delivery

David Weitz

This talk will describe new ways of making liposomes for drug delivery. Using an inverse-emulsion method, asymmetric liposomes are created. Each leaflet is formed separately enabling the formation of asymmetric structures, with different constituents in each. In addition, the outer leaflet can be coated with a polysaccharide, modifying its properties and providing a potential for improved targeting of the liposomes by matching specific molecules in the glycocylax of cells. These asymmetric liposomes are shown to work better for delivery of drugs, both for uptake and for endosomal release. This talk is dedicated to Bob Prud'homme who is a pioneer in liposome-based drug delivery.

40b Active Microrheology Method to Guide Mucoactive Agent Development

Gerald Fuller

Mucus viscoelasticity is determined by the underlying microstructure formed by the mucin polymer network, which is regulated by several factors including polymer concentration, density of covalent and physical bonds, and hydration. In airway diseases such as asthma and cystic fibrosis, mucus hypersecretion and dehydration results in thick, solid-like mucus that impairs transport, leading to mucus accumulation and blockage of the airway. Mucolytic drugs have been developed to reduce the viscoelasticity of the mucus gel. Reducing agents break down the disulfide bonds between mucins via thiol exchange chemistry. The most commonly used reducing agent drug is N-acetylcysteine; however, its clinical efficacy has been found to be limited due to low activity and a side-effect of bronchoconstriction. Novel reducing agents that are well-tolerated by human patients and effective at breaking down mucin crosslinks are in the process of being developed. Improved understanding of reducing agents effect on mucus material properties is needed to aid in the design of such drugs. Mucus-inspired hydrogels that capture aspects of the dynamic mucin network can help to elucidate the mechanisms of mucus rheology regulation in health and disease. We have demonstrated a magnetic microwire rheometer (MMWR) to measure the changes in material properties of dynamic hydrogels in response to agents that modulate disulfide-crosslinking. In this work, we use the MMWR to compare multiple reducing agents and characterize the degradation of disulfide crosslinked hydrogels as these agents diffuse through the network. While these reducing agents have been investigated via macrorheology methods, we show that the diffusion of these reagents, rather than mixing, into the hydrogel reveals key differences in degradation mechanisms that could have major implications for drug-delivery and efficacy *in vivo*.

40c Cellular Nanoparticles for Biological Neutralization

Liangfang Zhang

Biological neutralization is to deploy therapeutic agents to bind with harmful molecules or infectious pathogens, block their bioactivity, and thus prevent them from causing the diseases. The cell membrane-coated nanoparticles (namely 'cellular nanoparticles' or 'CNPs') offer a unique solution to address the challenges facing current biological neutralization technologies. Leveraging the fact that harmful agents need to interact with host cells for their bioactivity, the cellular nanoparticles function as cell decoys to intercept these harmful agents and divert them away from the intended cellular targets without a need to know the molecular structure of the agents. A shift of focus from the causative agents to the host cells make function-driven and broad-spectrum neutralization solutions possible. Herein, I will report on the latest development and scalable manufacturing of cellular nanoparticles for broad-spectrum neutralization against numerous pathological agents including bacterial toxins, inflammatory cytokines, and viruses.

40d Standing on the Shoulders of Flash Nanoprecipitation: Sequential Nanoprecipitation for the Continuous Manufacturing of Nanoparticles and Microparticles

Nathalie M. Pinkerton, Nouha El Amri, Rachel Pollard, Parker Lewis, Thomas Belinky, Allie Karakosta, Ashley Han

Sequential NanoPrecipitation (SNaP) is a nascent controlled precipitation process for the tunable formation of polymeric particles for drug delivery and bioimaging. While SNaP relies on the same self-assembly principles as one-step Flash NanoPrecipitation, SNaP is a two-step assembly process in which the particle core assembly is initiated during a first mixing step followed by particle stabilization in a second mixing step. Decoupling the particle assembly steps improves control over the particle structure and expands the attainable particle size range to include microparticles. Using a novel sequential mixer design, we attain unprecedented control over a key process variable: the delay time between the formation of the particle core and the particle stabilization. We demonstrate that particle size scales with delay time in agreement with Smoluchowski's model of diffusion-limited particle growth. This enables us to synthesize particles ranging from 50 nm to 5 μm in diameter and to encapsulate small molecule active pharmaceutical ingredients (APIs) and fluorophores with high encapsulation efficiencies via SNaP. Furthermore, SNaP can uniformly assemble components with different assembly timescales, for example, inorganic nanocrystals, polymer coils and small molecules, by affording control over the order of component addition and the delay time between additions. The novel SNaP process will unlock new particle architectures and open new avenues for nanoparticle and microparticle manufacturing in industry.

In honor of Prof. Bob Prud'homme's illustrious and impactful career as a curious scientist, engineer and mentor.

40e Process Variables That Control of the Morphology of Lamellar Crystals through Bending Elasticity

Maria Santore

Ultra-thin materials are of great interest in the creation of flexible electronics, conformal coatings, and materials that could be used to repair the body. Control of crystallization and the role of processing is also critically important in the pharmaceutical industry. We have been exploring how nanometrically thin fluid sheets such as phospholipid lamella enable morphological control of 2D (ultrathin) crystals that nucleate and grow within the lamellae themselves, for instance achieving thicknesses of only a few nanometers. Ultrathin crystals are mechanically solid and rigid to shear, they often prefer to be flat rather than curved due to molecular symmetry across a bilayer, and when they bend they tend to do so only cylindrically. When forced to nucleate and grow on spherical templates for instance a droplet surface, they experience prohibitive strain energy cost as a result of their inability to shear or stretch. As a result, beyond a size that increases with the size of the droplet, the crystal morphology shifts from a compact to a striped shape as the crystal grows. We have discovered that crystals growing within fluid lamella of closed vesicles exhibit the opposite dependence on vesicle size, with crystals growing within larger, more gradually curved vesicles adopting morphologies with greater protrusions. We show here that while lamellar crystals, like those growing on droplets, do not tolerate finite Gaussian curvature, when incorporated into fluid membranes, the bending of the membrane plays an important role in the overall energy, thereby dictating morphology. Further, when nucleation and growth results from a thermal quench into a two phase region, thermal membrane stress plays an important role in setting the bending costs, and this stress is relaxed by water transport. Hence the cooling history translates to a vesicle-size dependent tension history that controls crystal morphology. Batch processing methods that regulate thermal stress enable small scale production of highly regular lamellar crystals of select regular shapes (flowers, stars, hexagons) that correlate with vesicle size, facilitating their separations

40f The Challenge of Cost-Effective, Scalable, and FDA Approvable Nanoparticle Processes and Production

Robert K. Prud'homme, Walid Saad, Suzanne M. D'Addio, Brian K. Wilson, Sai Nikhil Subraveti, Bumjun Kim, Nicholas Caggiano, Jason X. Liu, Navid Bizmark

The dramatic implementation of the COVID lipid nanoparticle (LNP) vaccines have demonstrated the ability to produce nanoparticle therapeutics at scale, in a cost-effective manner. Mike McDermott, the President of Pfizer Global Supply, in the CNN interview where he took the press through the Pfizer production process stated: **"the impingement jet mixer makes it possible."** The confined impinging jet mixers are the technology developed in our laboratory. We will discuss the basis for the technology. The application of the technology for oral therapeutics will emphasize our work supported by the Bill and Melinda Gates Foundation on drugs for global health. The key question for the future is how will this technology be implemented for other therapeutic classes including high value small molecule drugs and targeted nucleotide therapeutics?

AUTHOR INDEX

AIChE
120 Wall Street, FL 23
New York, NY 10005-4020

ISBN 979-8-3313-1662-4

CURRAN ASSOCIATES INC.
proceedings.com